150
FOOD SCIENCE
QUESTIONS
ANSWERED

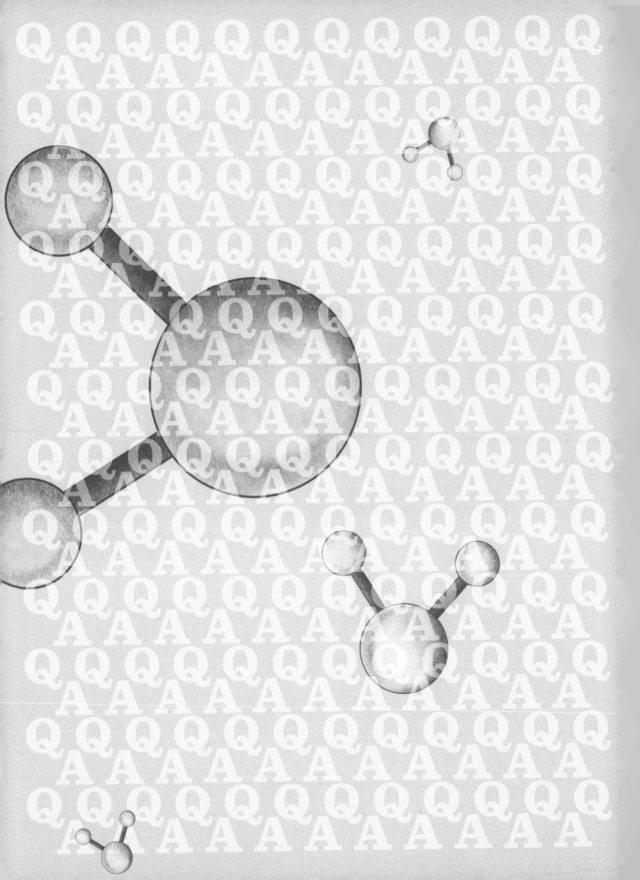

150
FOOD SCIENCE
QUESTIONS
ANSWERED

COOK SMARTER, COOK BETTER

BRYAN LE

ROCKRIDGE
PRESS

For general information on our other products and services or to obtain technical support, please contact our Customer Care Department within the United States at (866) 744-2665, or outside the United States at (510) 253-0500.

Rockridge Press publishes its books in a variety of electronic and print formats. Some content that appears in print may not be available in electronic books, and vice versa.

TRADEMARKS: Rockridge Press and the Rockridge Press logo are trademarks or registered trademarks of Callisto Media Inc. and/or its affiliates, in the United States and other countries, and may not be used without written permission. All other trademarks are the property of their respective owners. Rockridge Press is not associated with any product or vendor mentioned in this book.

Interior and Cover Designer: Amanda Kirk

Art Producer: Janice Ackerman

Editor: Pam Kingsley

Production Editor: Andrew Yackira

Illustrations: © 2020 Claire McCracken

Photography: Shutterstock: p. x; 5, 9, 11, 15, 16, 17, 24, 26, 29, 31, 33, 34, 36, 39, 41, 46, 48, 57, 61, 63, 65, 68, 69, 70, 72, 74, 76, 78, 83, 85, 86, 88, 90, 92, 94, 95, 99, 103, 104, 106, 107, 108, 110-111, 123, 128, 131, 133, 134, 135, 136, 137, 138, 140, 145, 147, 149, 150, 152, 155, 165, 167, 168, 169, 172, 174; istock: p. 67, 116.

ISBN: Print 978-1-64611-833-5 | eBook 978-1-64611-834-2

R0

TO MY WIFE, YVONNE,

who teaches me
to laugh every day

Contents

CHAPTER 4

EGGS AND DAIRY 71

FOOD SAFETY AND STORAGE 141

Introduction

AS THE OLD SAYING GOES, "EVERYONE'S GOTTA EAT!" Regardless of who you are, where you come from, what you do, and what you believe in, you have to eat (and several times a day, no less).

The truth is, I didn't think too much about food growing up. Food was never that important to me, and I wasn't a picky eater. To me, eating was just a way to get the energy I needed to read about science and run experiments in my garage laboratory. I was that nerdy kid who spent his spare hours collecting bugs, mixing chemicals, and growing mold just for fun. When I went off to college, naturally, I studied chemistry. And while I loved going to class and learning about reactions, chemical structures, and thermodynamics, I ended up graduating without knowing what my next step was. I was so desperate for a direction that I spent six months walking the 2,000 miles from Los Angeles to New Orleans just to clear my head.

When I got back to L.A., I reconnected with the woman who would eventually become my wife, and over the next couple years, I tried my hand at a number of things in an effort to figure out what I really wanted to do professionally. During this time, my future wife started introducing me to different cuisines, first through the restaurants we frequented and then in our own kitchen. We learned to cook together (she's still the better cook), and I began to value food beyond its role as fuel for my activities. My then-girlfriend's love of food spread to me, and we spent many happy moments cooking and eating together.

At some point around that time, I stopped by a university library and happened upon a research journal about the chemistry of food and flavors. My first thought was, "You can study that?!" I picked up a copy of the journal and started reading about the science and technology behind what makes food taste great, which was something I'd never considered as an academic field of study before. That journal flipped a switch in me, reigniting my desire to learn and discover, so I applied to a graduate-level food science program.

During the course of my food science studies, I began meeting other people who loved food so much they were willing to spend years researching it. Their passion was infectious, and I began reading more and more about science, technology, and food. In the graduate program, I was able to connect my knowledge and understanding of chemistry with something that gave people so much joy and comfort—food. I was amazed to learn that by controlling heat, moisture, acidity, and salt content, I could almost magically transform the way flavors are produced in a searing steak or stir-fry. I was blown away by the complex chemical reactions behind the aromas of freshly baked bread. And I loved that I could combine ingredients in precise proportions to produce something wonderful, exciting, and tasty for people to enjoy, much in the same way that I could combine chemicals in the laboratory (okay, except for the tasty part). Understanding that there were direct, scientific ways in which I could deliberately change and manipulate food to make better dishes helped me become a better cook. And diving deeper into the science behind food and cooking revealed dimensions of food that I had not yet explored, giving me a better appreciation for what sits in front of me at each meal. My hope is that this introductory guide to the fascinating science of food will give you that same inspiration and appreciation.

CHAPTER 1

Cooking Basics

COOKING CAN BE KIND OF INTIMIDATING. When I first decided to learn more about cooking and food, I wasn't quite sure where to begin. I had lots of questions, so I started with the basics, taking the fundamental principles of chemistry and physics that I'd learned in school and applying them to cooking. In this chapter, you'll find a series of broad questions that will help you explore the science behind making food and debunk a few misconceptions along the way.

What Is Cooking?

THE SCIENCE Cooking is the process of taking raw food and turning it into something edible, nourishing, and/or tasty. Many of the processes that occur during cooking are based on principles found in biology, chemistry, and physics. For example, much of our experience of flavor comes from the chemical reactions that occur between amino acids and sugars at high temperatures. And the softening of foods like pasta, potatoes, and beans is a result of the interaction of hot water molecules with starch granules. Just as it can help explain many processes found in the natural world, scientific theory can help shed light on the reasoning behind certain cooking techniques and advice found in cooking lore.

Let's start with the basics. In order to facilitate the chemical and physical transformations that are involved in cooking, energy needs to be brought in from an external energy source. The main energy source humans have used for eons is heat. While fire was the only heat source for early humans, modern humans have several sources of heat available that involve heat transfer through conduction, convection, and/or radiation. Conduction is the most common type of heat transfer found in the kitchen and involves the transfer of heat from one material or surface to another—for example, heating a steak in a hot pan or grill. Convection is the transfer of heat through a gas or liquid—in cooking, typically water, steam, or air. Examples of convection at work in the kitchen include hard-boiling eggs

HARNESSING HEAT
GIVES HUMANS ACCESS TO A WIDE RANGE OF POTENTIAL FOOD MATERIALS FROM NATURE

and steaming vegetables, and the movement of hot air in a convection oven. Radiation is heat transfer through particles of light. A common form of radiation heat transfer is found in the microwave oven, where microwave radiation is used to quickly raise the temperature of water-rich food.

Harnessing heat gives humans access to a wide range of potential food materials from nature. The heat used in cooking not only helps develop flavors and improve texture, but also deactivates enzymes that cause spoilage and degrade nutrients, breaks down toxins, kills off microbial pathogens, and unlocks certain nutrients in raw vegetable and animal tissues that are otherwise inaccessible. Harvard anthropologist and MacArthur Fellow Richard Wrangham has proposed that being able to access nutrients from animal tissues enabled early humans to evolve larger brains. Cooking food could very well be one of those special powers that make us unique as humans.

KITCHEN TAKEAWAY Cooking was one of the first ways humans experimented with and explored our world. As humans have evolved, so has cooking—we are past the point of just throwing meat on the fire and calling it a day. By understanding what cooking means and effectively applying the scientific principles involved, anyone can create nutritious, flavorful dishes that satisfy the heart, body, and soul.

Q IS ONE KIND OF MATERIAL BEST FOR COOKWARE?

THE ANSWER No

THE SCIENCE Cookware can be made from a variety of different materials—most commonly iron, copper, and aluminum. Each metal has its own unique set of physical and chemical properties that make it work well for specific applications. Iron, for example, is a tough, dense metal that is resistant to impact. Cookware made from iron (e.g., stainless steel, carbon steel, cast iron) is durable, and its high density yields better heat retention. The main drawback of using iron cookware, however, is that, because of its high density, it requires more time to heat up. Iron is also a relatively reactive metal, so cast-iron and carbon-steel cookware will begin to corrode if improperly maintained. Stainless steel is an exception to this because it contains chromium metal in its alloy formulation, which makes this material resistant to corrosion by water, air, acids, and alkaline ingredients. Enameled cast iron is also resistant to corrosion because of the glazing applied to its surface.

Aluminum and copper cookware present unique advantages and disadvantages over their iron counterparts. Aluminum and copper have some of the highest heat transfer rates of all metals, which allow for quick heating, even heat distribution, and rapid decrease in temperature when removed from the heat source. Aluminum is also quite lightweight, but has relatively poor heat retention, which means that the temperature of a hot aluminum pan will drop when cold food is added. The high density of copper allows it to efficiently retain and distribute heat in much the same way iron does, while being much quicker to heat up and cool down. A common disadvantage of both copper and aluminum, though, is that their malleability makes them prone to denting and warping. Also, both are reactive and can leach metals into acidic foods.

Fully clad (a.k.a. all-clad) stainless cookware works around both issues: the low heat transfer rate of stainless steel and the reactivity of copper and aluminum metals. The bottom of fully clad cookware is made up of an aluminum or copper core sheathed in stainless steel. The stainless steel protects the cookware from corrosion caused by air, acid, and moisture, and provides strength and durability, while the copper or aluminum core increases heat transfer. The downside of fully clad cookware is that the extra layers of metal make these pots and pans heavier than comparably sized models made with 100 percent stainless steel or other metals.

KITCHEN TAKEAWAY No single material is best for all cooking applications. Iron and steel cookware are good for applications that require high heat and good heat retention, like searing, roasting, and frying. Aluminum cookware is great for quick cooking that doesn't require a lot of heat management, like boiling, steaming, or roasting. Copper is best reserved for cooking delicate foods that require the ability to quickly change temperatures, like seafood, sauces, caramel, and chocolate.

NO SINGLE MATERIAL
IS BEST FOR ALL
COOKING APPLICATIONS

WHAT HAPPENS WHEN YOU SEASON A CAST-IRON PAN?

THE SCIENCE The bare metal of a cast-iron pan can easily corrode and stick to food during cooking, but seasoning the surface creates a rust-resistant, nonstick coating. The process usually involves washing a new pan with hot, soapy water to remove any residue left from its manufacture, drying it completely, adding a thin coat of oil to the entire pan (including the exterior and handle, if those parts are also made from cast iron), and heating the pan in a 350°F/177°C oven for about an hour. This heating causes unsaturated fat molecules to react with the oxygen in the air to form peroxides (similar to hydrogen peroxide). These peroxides further react with neighboring unsaturated oil molecules to form bridges in a chain reaction that hardens the oil into a thin, water-repelling polymer coating. During this process, the oil coating on the pan is essentially cured in the same way that vegetable oil–based paint varnishes are cured. Repeated cycles of applying oil and heating the pan for 30 to 60 minutes will thicken the coating, which is why it's recommended to season cast-iron pans three or four times before you use them.

KITCHEN TAKEAWAY Before seasoning your cast-iron pan, make sure it's washed and dried thoroughly, then heat it in the oven for a few minutes to get rid of any residual water before applying the oil. Use oils that have high concentrations of polyunsaturated fats, such as corn, sunflower, flaxseed, olive, and grapeseed oil (coconut oil doesn't work). Although a well-seasoned cast-iron pan can handle occasional acidic ingredients, don't put them in your pan on a regular basis. Also, make sure you don't leave a seasoned pan to soak in water. While the coating is resistant to water, microscopic imperfections can let water and air seep in and rust the pan.

Is Steam Really Hotter Than Boiling Water?

THE ANSWER Yes

THE SCIENCE Water is a molecule made up of the two chemical elements hydrogen and oxygen, with two hydrogen atoms attached to one oxygen atom in the shape of a Mickey Mouse head (the ears are the hydrogens). Hydrogen is the smallest atom in the universe, with only one lonely electron orbiting its nucleus, which consists of a single positively charged proton. Oxygen, on the other hand, has eight electrons revolving around its nucleus of eight protons. When these three atoms are combined, the hydrogens' electrons are strongly attracted to oxygen's large nucleus. Those two extra negatively charged electrons orbiting the oxygen give the oxygen in the molecule a slightly negative charge. The two hydrogen atoms, which have basically lost their electrons, end up with a slightly positive charge.

Just like in dating, opposites attract in the world of chemical bonding and atomic physics. So, if two water molecules are next to each other, the slightly

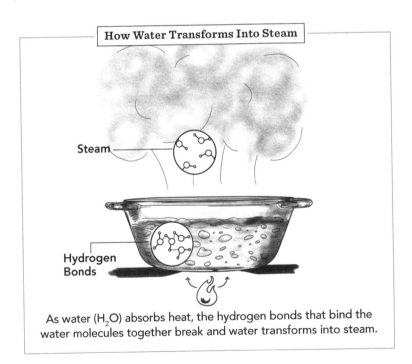

How Water Transforms Into Steam

Steam

Hydrogen Bonds

As water (H_2O) absorbs heat, the hydrogen bonds that bind the water molecules together break and water transforms into steam.

negatively charged oxygen of one water molecule will be attracted to the slightly positively charged hydrogen of a neighboring water molecule. This attraction creates a momentary "stickiness" between the two molecules, called a hydrogen bond. In any given amount of water, there are a huge number of these hydrogen bonds constantly forming, breaking, and re-forming as all the water molecules move around one another. This strong network of constantly re-forming hydrogen bonds is the reason why water is liquid at room temperature on our planet.

When water is heated up, the water molecules gain energy and start moving faster, which causes the hydrogen bonds to break up more often than they re-form. This process accelerates until water reaches its boiling point at 212°F/100°C, where there is so much movement between the water molecules that all of the hydrogen bonds break and there is nothing to keep the water molecules stuck together, at which point the water converts to steam, the gaseous state of water.

Boiling water is always stuck with this 212°F/100°C limit (at least, at normal sea-level pressures) because every high-energy liquid water molecule that reaches that temperature automatically transforms into steam and leaves the pot or pan. Steam, on the other hand, has no other form to transform into as its temperature increases. If you keep adding energy in the form of heat to steam, the gaseous water molecules just keep moving faster and faster, and the temperature climbs higher and higher. However, the practical temperature limit to steam in the kitchen is 212°F/100°C because no additional energy is added after steam escapes from a boiling pot of water. That being said, if the pot's lid is locked on with an airtight seal, like on a pressure cooker, the temperature of the steam can reach 250°F/121°C.

Steam does contain more heat energy than boiling water, because a lot of energy is needed to break the hydrogen bonds; therefore, heat generated from steam is transferred faster to foods than heat generated from boiling water.

KITCHEN TAKEAWAY Steam foods when you want them to remain moist but not waterlogged. As a tip, make sure to keep the lid tight on the steamer to prevent excess heat and moisture from escaping. And remember, while steam can exceed the boiling point of water, the water in food is still limited by that boiling point temperature, so the maximum temperature you can achieve inside a food remains 212°F/100°C. Be careful when removing the lid from pots or releasing steam from a pressure cooker, as the steam can scald exposed skin in an instant just as badly as boiling water can.

Q DOES COLD WATER COME TO A BOIL FASTER THAN WARM WATER?

THE ANSWER No

THE SCIENCE From a thermodynamic perspective, you don't have to add as much heat to a warm pot of water to get it to reach boiling, so it'll take less time. So why do some people think cold water boils faster than warm water? Some scientists speculate that this mistaken belief comes from a misperception about how water is heated. People assume that water should heat linearly, that the time it takes to go from 50°F to 100°F/10°C to 38°C should be the same as going from 100°F to 150°F/38°C to 66°C. In actuality, while heat initially transfers at a faster rate into cold water, the transfer begins to slow down as the water becomes hotter and begins to evaporate and lose heat through radiation (to the pot its being boiled in, for instance). As a result, it takes longer to heat a pot of warm water 25°F higher than it does to heat a pot of cold water 25°F higher. That said, the pot of cold water still must climb many more degrees to get to the boiling point, compared with the pot of warm water.

KITCHEN TAKEAWAY Start with a pot of warm water from the tap if your goal is to get to boiling as fast as possible. However, one reason to start with cold water would be to ensure even heat distribution in the food you're cooking, such as potatoes (see Is There Really a Difference Between Starting Potatoes in Cold Water Versus Adding Them to Boiling Water?, page 109).

Why Does Water Boil at Less Than 212°F/100°C at Higher Altitudes?

THE SCIENCE If you've ever left a cup of water on a table for a while, you've probably discovered that the water will evaporate over time. This happens because some of the liquid water molecules have just enough energy at room temperature to break apart from the other water molecules to escape as a vapor or gas. The process is slow because there isn't a whole lot of energy at room temperature, but as ambient temperatures rise, the evaporation rate increases—think about how quickly a puddle will evaporate on a hot summer day.

Boiling is the same process as evaporation, except that the energy from the heat makes the liquid molecules want to become gas, though the process will take some time, as you may have noticed if you've ever put a pot on to boil and then forgotten about it for half an hour. The reason a pot of water takes time to boil is that the air molecules of our atmosphere take up space right at the surface of the pot and weigh down those water molecules that want to vaporize. In scientific terms, these molecules exert air pressure. If we placed a cup of water in the vacuum of space where there are no air molecules and practically no air pressure, that water would quickly vaporize into gas even at freezing temperatures. But on Earth, our thick atmosphere of air molecules puts pressure on water molecules and slows down the vaporizing process.

At higher altitudes, there are fewer air molecules to weigh down the water molecules in a boiling pot, so it takes less energy for the water molecules to turn into gas, meaning that water can come to a boil (start converting into a gas) at a temperature lower than 212°F/100°C.

KITCHEN TAKEAWAY Though it may seem counterintuitive, high-altitude cooking that involves heat transfer through water or steam (which includes pretty much every kind of cooking) will take longer because it is taking place at a lower temperature. Think of the time difference between cooking a leg of lamb at 300°F/149°C versus at 375°F/~191°C. For every 500 feet/152 meters in elevation about sea level, water's boiling point decreases by about 1°F/~0.5°C. At 3,000 feet/914 meters above sea level, pasta requires 25 to 50 percent more time to reach al dente than it does at sea level. When cooking at 5,000 feet/1,524 meters above sea level, cook times for meat should be increased by 25 percent (since most recipes are written for sea level altitudes). Also, foods tend to dry out faster at high altitudes, so a tight-fitting lid should be used when braising meats or vegetables or boiling water. When baking at high altitudes, oven temperatures should be set 15°F to 25°F/8°C to 14°C higher than indicated in most cookbooks, and baking times should be decreased by 20 to 30 percent to compensate for the hotter oven temperatures.

DOES ADDING OIL TO THE COOKING WATER KEEP PASTA FROM STICKING?

THE ANSWER No

THE SCIENCE Pasta is composed largely of starch. Starch is a very large molecule made up of long, branched strings of sugar molecules. During the process of cooking pasta, these starch molecules absorb water like little sponges. Some of the surface starches will become incredibly sticky as a result, in much the same way that sugar dissolved in water becomes sticky. These starch molecules will dissolve into the water during cooking, but if there isn't enough water in the pot, and the water becomes saturated with starch, the pasta will remain sticky.

It's commonly thought that adding oil to the cooking water will keep the pasta from sticking to itself. The trouble with this theory is that oil is less dense than water and floats on the surface of the water, leaving the pasta uncoated while it cooks. However, the oil will coat the pasta as it drains, which will prevent your noodles from soaking up any water-based sauce you add to it.

KITCHEN TAKEAWAY To keep pasta from sticking together as it cooks, use a large pot with at least 4 quarts of water per pound of dried pasta, and stir the pasta frequently during cooking.

OIL IS LESS DENSE THAN WATER
AND FLOATS ON THE SURFACE OF THE WATER, LEAVING THE PASTA UNCOATED WHILE IT COOKS

Do Meats and Vegetables Cook in the Same Way?

THE ANSWER No

THE SCIENCE The way a food cooks depends on its chemical composition and how the compounds in the food respond to heat. While both meats and vegetables are made of mostly water, meats contain large amounts of protein, amino acids, and fat, whereas vegetables are composed primarily of complex carbohydrates like starches and fiber. When heat is applied to raw meat, the internal temperature of the meat rises until the cell walls of the muscle tissues fall apart and the molecular configurations of the meat proteins break down (this is known as denaturing). As a result of this process, the meat softens and releases water (which we know as meat "juice").

Collagen, the main protein that supports the connective tissue in meat, will also shrink, soften, and expel water during cooking. As the meat's temperature continues to climb, amino acids and sugars in the meat juices react together and brown to form what we think of as meaty flavors and aromas (see Why Do Foods Brown? The Maillard Reaction, page 12).

When vegetables are cooked, water is released from their carbohydrate-rich structure. The starches begin to swell and unfold as water migrates from different tissues, causing the cell walls to burst open. The carbohydrates, especially starch, break down and form complex and simple sugars. Browning reactions can also occur with vegetables, but since they tend to contain more sugars than amino acids, the flavor profiles of vegetables are quite different from those of meats.

KITCHEN TAKEAWAY While meats and vegetables cook differently, one thing they have in common is that they're both made mostly of water. Being able to control water and its properties, largely through heat, is an important feature of cooking.

Why Do Foods Brown?
The Maillard Reaction

THE SCIENCE Browning is a miraculous chemical reaction that gives food its rich flavors and delightful aromas. Food browning reactions are important in nearly all methods of high-heat cooking, like roasting, searing, and baking. There are two different processes, each dependent on a different chemical reaction, that are responsible for browning in food: the Maillard reaction and caramelization (see page 13). In the Maillard reaction, as the food heats, certain amino acids and simple sugars on the surface of the food interact to create flavor compounds, each of which can further react with fats and other amino acids to create hundreds more flavor compounds. Many of these compounds absorb light strongly, which is why foods turn brown during this process. The process is named after Louis Camille Maillard, who, in 1912, first identified the relationship between amino acids and simple sugars that made this discovery possible.

Since the Maillard reaction is the basis for many enjoyable flavors, flavor companies recreate it to produce natural and artificial flavors. It has also been discovered that Maillard reactions involving different amino acids yield different flavor profiles. For example, one of the simplest ways to create a meaty flavor is to heat the amino

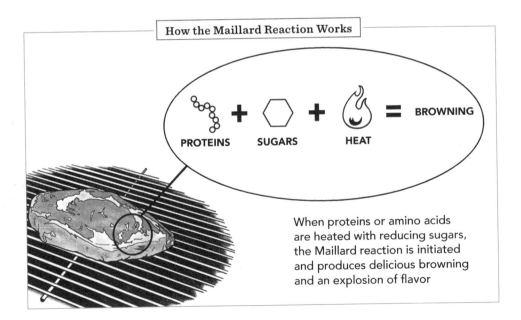

How the Maillard Reaction Works

PROTEINS + SUGARS + HEAT = BROWNING

When proteins or amino acids are heated with reducing sugars, the Maillard reaction is initiated and produces delicious browning and an explosion of flavor

acid cysteine with glucose for several hours. Other amino acids and sugars can be combined to form caramel, roasted coffee, chocolate, or vegetable broth flavors.

Foods containing combinations of amino acids and simple sugars that are ideal for the Maillard reaction include bread, onions, potatoes, and meats.

KITCHEN TAKEAWAY For the Maillard reaction to be triggered, food needs to be heated to between 280°F and 300°F/138°C and 149°C, which is why recipes often call for oven temperatures of 350°F/177°C or higher. High-heat cooking like searing or roasting brings surfaces to Maillard temperatures and beyond very quickly so foods can cook fast without burning. The Maillard reaction can still take place at temperatures below 280°F/138°C, but it will happen much more slowly.

Why Do Foods Brown?
Caramelization

THE SCIENCE The Maillard reaction is distinct from caramelization in that it requires the interaction of sugar and amino acids or proteins, whereas caramelization is all about sugar. Caramelization occurs when sugars are heated until they break apart into new flavor molecules, a process key to producing the colors and flavors we love so much in peanut brittle, crème brûlée, and caramelized onions, for instance. The molecules most responsible for the flavors we associate with caramelization are maltol (caramel flavor) and furan (nutty flavor). Different sugars caramelize at different temperatures. Fructose, which is mostly found in fruits, onions, corn syrup, honey, and soft drinks, starts to caramelize at 230°F/110°C. Glucose and sucrose (or table sugar) begin to caramelize at 320°F/160°C. Both Maillard browning and caramelization can happen at the same time if there's also a source of protein or amino acids present.

KITCHEN TAKEAWAY Caramelization is important for creating nutty and caramel flavors in sugar-rich foods and desserts. Caramelization can be sped up by decrease the pH to below 7 by adding an acid like lemon juice or tartaric acid (cream of tartar). Acidity helps catalyze the reaction of sucrose (table sugar) with water to form glucose and fructose, which further decompose into the compounds responsible for caramel flavor.

Can You Increase Browning?

THE ANSWER Yes

THE SCIENCE The Maillard reaction is highly dependent on pH; alkaline foods will brown more deeply than acidic foods will. However, as the Maillard reaction takes place, acidic flavor compounds are generated, which cause the reaction (and browning) to slow as the acidity level rises. The reason for this effect is that as the overall pH decreases, amino acids become less available to react with sugars. But, if this acidity is removed, the Maillard reaction will continue to churn out more flavorful compounds. The easiest way to reduce the acidity of a food is to add baking soda, an alkali, to a recipe, which will increase the pH. This is why you will often see baking recipes that call for both baking powder and baking soda—the baking powder helps with leavening, and the baking soda is there to both promote browning and offer extra leavening support.

Another way to accelerate the Maillard reaction is to add free amino acids and certain sugars to the food. The most kitchen-friendly source of free amino acids is egg whites. Baking recipes that call for lightly brushing dough with beaten egg whites rely on the Maillard reaction to create a tasty, brown crust. Adding sugars, such as glucose, fructose, and lactose, can also promote the Maillard reaction; common sources are corn syrup, soda, milk, honey, and agave nectar.

Temperature is an important factor for the Maillard reaction. As in practically every other chemical reaction, increasing temperature can increase the rate at which the Maillard reaction occurs. One simple way to speed up Maillard browning is to increase the temperature, but this only works up to a point. According to Nathan Myhrvold, founder of The Cooking Lab and co-author of the award-winning book *Modernist Cuisine*, optimal temperatures for the Maillard reaction are between 280°F and 355°F/138°C and 179°C. Searing brings the surface temperature of a steak to Maillard temperatures very quickly, which allows browning to occur almost instantly. However, charring starts to occur at temperatures above 355°F/179°C, which produces bitter compounds and is the reason why searing for too long isn't recommended. It is also important to note that the presence of water (as in oven braising, for instance) can also interfere with the temperature of the food, keeping it at the boiling point of water, 212°F/100°C, regardless of the temperature of its surroundings.

KITCHEN TAKEAWAY According to Samin Nosrat, author of the James Beard Award–winning cookbook *Salt, Fat, Acid, Heat,* if you want to brown something quickly, salt it after the food has developed a crisp edge, or salt it ahead of time and pat the food dry with a paper towel before it goes in the pan. Salt draws moisture to the surface of a food, and that liquid can impede the Maillard reaction. Roast and sauté vegetables and meats with plenty of space in between pieces to let Maillard-inhibiting steam escape from the pan. A sprinkle of baking soda directly on meat and vegetables helps to accelerate browning by making amino acids more available to react.

DOES ALCOHOL BURN OFF WHEN YOU COOK IT?

THE ANSWER Sort of

THE SCIENCE In recipes that call for alcohol, such as the wine in chicken Marsala, for instance, you are often instructed to add the alcohol and simmer the recipe for a few minutes to "burn off the alcohol." However, a USDA study found that even if you flambé a dish, which involves igniting it, only 25 percent of the alcohol

 is evaporated in the process. The only way to evaporate a significant portion of the alcohol in a dish is through long simmering; the same USDA study found that 2½ hours of simmering will make all but 4 to 6 percent of the original alcohol evaporate from a recipe.

KITCHEN TAKEAWAY If your goal is an alcohol-free dish, it's best to start with an alcohol-free wine or beer or another liquid altogether.

Does the Quality of a Wine Matter When You're Cooking with It?

THE ANSWER Not really

THE SCIENCE You've heard the adage, "Never cook with a wine you wouldn't drink." But is the quality of cooking wine really that important? In 2007, Julia Moskin at the *New York Times* tested both low- and high-quality wines cooked in different dishes and found that they produced equally tasty food. So, what gives? The cooking process evaporates some of the alcohol (the longer the dish cooks, the more alcohol is evaporated), as well as many of the flavor compounds that give wine its distinctive flavor. What is left of the wine after cooking are compounds that are common in all wines. One important remnant is the slowly decaying remains of the dead yeasts that once lived in and fermented the wine. The yeasts contain many compounds that elicit umami and kokumi flavors (see Is Umami Really a Thing?, page 34), including glutamates, inosinates, guanylates, and glutathione, which add savory flavor to the dish. Other compounds in wine that do not evaporate during cooking are tartaric acid, glycerol, tannin, sugar, and trace minerals. Tartaric acid adds acidity to a recipe, while glycerol and sugar add sweetness. All of these components can help balance a dish. And while there are certainly variations between these compounds depending on the exact wine you use, along with water and ethanol, they make up 99.5 percent of all wines. In comparison, the volatile flavor compounds in wine (the notes that indicate you're drinking "good" wine) make up only 0.5 percent of its composition.

KITCHEN TAKEAWAY Use a cheap wine to cook with, and save the finer bottles for drinking. However, one thing to watch out for is the level of tannins in a red wine, which can add bitterness and astringency to a dish. Err on the side of a smoother, more medium-bodied red wine. Sweet wines can also get concentrated during cooking, which can throw off the flavor balance in a dish.

IS OIL REALLY NECESSARY IN COOKING?

THE ANSWER It helps

THE SCIENCE Although water is a convenient and readily available liquid that can be used to cook all sorts of foods, its temperature is limited to its boiling point (see Is Steam Really Hotter Than Water?, page 6), and a lot of interesting food chemistry happens above that temperature, namely Maillard flavor reactions and crisping. That's where oils come in. These liquid fats can be heated to a much higher temperature than water and can transfer heat quickly to food. This allows the surface of a food to reach higher temperatures, which helps it cook faster, increase the rate of flavor production, and develop a crunchy exterior. Many flavor molecules are also fat-soluble, and oil helps to concentrate these flavors. Not to mention, oils are delicious on their own.

Oil has other properties that help the cooking process, too. It can help redistribute heat on any cooking surface that heats unevenly, and it can also prevent food from sticking to pots and pans. Food sticks to cookware by forming a chemical bond between the sulfur atoms of proteins in the food and the metal surface when heated. Oil creates a barrier between these proteins and the heated surface that keeps this chemical bond from forming.

KITCHEN TAKEAWAY When you cook with oil, make sure to cover the entire bottom of the pan with an even layer of oil; otherwise, heat won't properly transfer to all parts of the food you are cooking, leaving some parts undercooked or underflavored.

Q DOES THE TYPE OF OIL I USE MATTER?

THE ANSWER Yes

THE SCIENCE When it comes to oils, which are just liquid fats, the science can get complex. In any given oil, you have some percentage of polyunsaturated fats, monounsaturated fats, and saturated fats. Whether a fat is saturated or unsaturated has to do with the structure of the fat molecule. Fats are made of triglycerides, which consist of three fatty acids connected to a glycerol molecule. Fatty acids are made of linear chains of carbon atoms of varying lengths. How those carbons are attached to each other determines if the fatty acid has saturated or unsaturated bonds. A saturated bond consists of all single bonds between two carbons, while an unsaturated bond has two or more bonds between two carbons in a molecule. Fatty acids with all or mostly all saturated bonds, like saturated or monounsaturated fatty acids, are more resistant to heat and air.

Fatty acids that have several unsaturated bonds, or polyunsaturated fats, are more likely to burn or become rancid when left out. The unsaturated bonds are the weak points in a fatty acid structure. Oils with high amounts of saturated or monounsaturated fats are more resistant to high heat than oils rich in polyunsaturated fats because those fats are more reactive to oxygen.

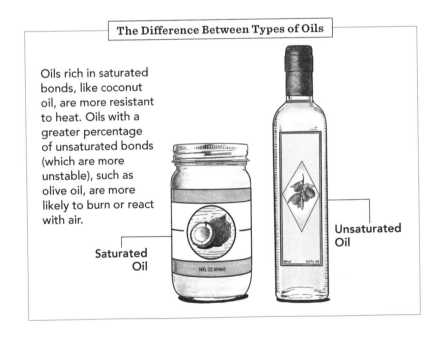

The Difference Between Types of Oils

Oils rich in saturated bonds, like coconut oil, are more resistant to heat. Oils with a greater percentage of unsaturated bonds (which are more unstable), such as olive oil, are more likely to burn or react with air.

Saturated Oil

Unsaturated Oil

14FL OZ (414ml)

THE FLAVOR OF AN OIL
COMES FROM THE FLAVOR COMPOUNDS EXTRACTED FROM THE HARVESTED PLANT MATERIAL ALONGSIDE THE OIL

The smoke point of an oil is the temperature at which it decomposes into smaller fragments, which then produce smoke as they vaporize. Coconut oil contains 98% saturated and monounsaturated fats and has a smoke point of 450°F/232°C, which makes it a good choice for deep frying. Extra-virgin olive oil is largely monounsaturated but does contain 15% polyunsaturated fats, yielding a lower smoke point that ranges from 320° to 375°F/160° to 190°C and making it better suited for comparably lower temperature cooking methods like sautéing or in uncooked preparations like dips and vinaigrettes.

Pure oil, whether it's high in saturated or unsaturated fat, has no taste at all. The flavor of an oil comes from the flavor compounds extracted from the harvested plant material alongside the oil, like robustly flavored olive oils or nut/seed oils like walnut, hazelnut, and sesame. To best preserve their flavor, these oils should be heated minimally or not at all, as their flavor molecules are susceptible to evaporation and oxidation when heated at high temperatures.

KITCHEN TAKEAWAY High-smoke-point oils with neutral flavors (avocado, coconut, peanut, vegetable, and corn oils) are best used for high-temperature applications like frying. Low-smoke-point oils (olive, sunflower, safflower, flaxseed, grapeseed, and unrefined coconut oils) are better for lower-temperature applications like sautéing and baking. Sesame seed and walnut oils are best used in preparations that require no cooking, like salad dressings, or stirred in right before serving to preserve their flavors.

WHAT IS AN EMULSION?

THE SCIENCE Normally, fats (usually oil) and water do not mix; instead, they form separate layers if they are poured into the same vessel. However, if oil and water are strongly whisked or mechanically beaten together, they can be forced to overcome that natural repulsion and form an homogeneous mixture known as an emulsion.

There are two types of emulsions: water in oil (w/o) and oil in water (o/w). In the case of a w/o emulsion, the whisking breaks the water (or water-containing liquid, like vinegar) into microscopic droplets that are evenly dispersed into the fat—such as the olive oil in a vinaigrette, for instance. In an o/w emulsion,

the opposite is true. Examples of o/w emulsions are mayonnaise, hollandaise sauce, butter, and cream.

One of the defining characteristics of an emulsion is the thickening that takes place as it comes together; this occurs because the larger, slower moving oil molecules interfere with the movement of the smaller, fast-moving water molecules, resulting in an increase in viscosity.

KITCHEN TAKEAWAY Patience is key when it comes to making emulsions like mayonnaise or hollandaise sauce. The microscopic droplets need time to disperse evenly when whisked, so be sure to slow things down and resist the urge to try shortcuts.

WHY ARE EMULSIONS PRONE TO BREAKING?

THE SCIENCE As mentioned above, when you make an emulsion, you are mechanically forcing water and oil to go together by whisking the dickens out of them. But this newly harmonious state is unstable. In the case of vinaigrette, the water-rich vinegar droplets

will start to attract to one another, then merge together to form larger and larger drops until they separate out entirely. The oil droplets will do the same with other oil droplets.

To slow down this separation (or prevent it entirely) and aid in the initial emulsification, compounds known as emulsifiers are added. The molecular structure of an emulsifier contains both hydrophobic (water-hating) and hydrophilic (water-loving) components. In the case of an oil-in-water emulsion whisked with an emulsifier, the oil droplets are coated with emulsifier molecules whose water-hating ends face inside toward the oil, while the water-loving parts face outward toward the water. In this way, the oil droplets become encased in oil-repelling shields that keep the oil droplets from separating out and aggregating and simultaneously allow water to surround each oil droplet.

For the home kitchen, common emulsifying agents are egg yolks, mustard, garlic paste, and butter. For more on specific emulsions, see Why Won't My Mayonnaise Thicken? (page 80) and Why Did My Hollandaise Sauce Break? (page 81).

KITCHEN TAKEAWAY If you are having a hard time getting your vinaigrette to thicken up, add a bit of mustard or mashed garlic to help the emulsion form.

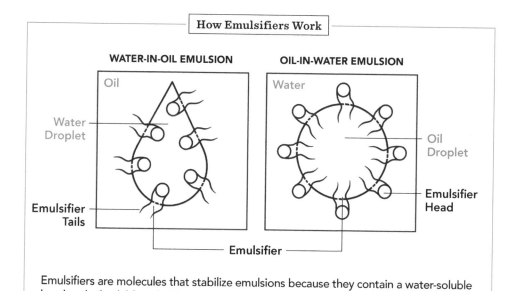

How Emulsifiers Work

WATER-IN-OIL EMULSION

Oil

Water Droplet

Emulsifier Tails

OIL-IN-WATER EMULSION

Water

Oil Droplet

Emulsifier Head

Emulsifier

Emulsifiers are molecules that stabilize emulsions because they contain a water-soluble head and oil-soluble tails. The water-soluble heads orient themselves inward toward the water droplet in water-in-oil emulsions and the oil-soluble tails orient themselves outward toward the surrounding oil. In an oil-in-water emulsion, the positions are reversed.

Does It Matter What Thickener I Use?

THE ANSWER Yes

THE SCIENCE There are many thickeners available for the modern cook. More traditional choices include cornstarch, all-purpose flour, rice flour, potato starch, arrowroot powder, and tapioca starch, but there are also alternative thickeners like coconut flour, chia seeds, and psyllium husks. And then there's the high-potency thickeners guar gum, xanthan gum, gellan gum, and agar agar powder. With so many thickeners to choose from, how do you know what's best? Let's start with some basics about what exactly thickeners are and how they work.

Most thickeners are some form of complex carbohydrate, usually derived from a type of starch or fiber. These carbohydrates are composed of hundreds, sometimes thousands, of chemically bonded sugar molecules in long chains or networks commonly referred to as polysaccharides. Normally, when sugar is added to water, it dissolves because the water molecules easily surround the crystals and form what is known as a hydration shell. Water also forms a hydration shell around the polysaccharides in thickeners, but the immense size of these carbohydrates prevents them from easily dissolving.

When thickeners are heated in water, they unravel their long sugar chains once they reach a specific temperature called their gelatinization temperature. This is the point

MOST THICKENERS ARE
SOME FORM OF COMPLEX CARBOHYDRATE, USUALLY DERIVED FROM A TYPE OF STARCH OR FIBER

at which hydrogen bonds between the sugar chains break and large numbers of water molecules take up residence in the deeper crevices of the polysaccharide network, causing the polysaccharide molecules to swell in size. At the same time, these polysaccharides latch onto neighboring polysaccharides and create super large networks and chains. The result is the thickening effect we associate with thickeners—these hydrated polysaccharide networks slow down the movement of water molecules and dramatically increase the viscosity of the solution.

The chemical composition of a thickener affects its rate of water uptake and network strength, which affects the viscosity of the recipe you are adding it to. The food source from which the thickener is produced will affect its polysaccharide composition, including the types of sugar molecules that make up its structure, the structural arrangement of its bonds, and the size of its molecular network.

KITCHEN TAKEAWAY Wheat flour gives an opaque and matte appearance to dishes at 2 tablespoons per cup of liquid, and it works best when cooked with oil or butter in a 1:1 ratio to remove the raw flour flavor and evenly disperse the flour so it doesn't clump when the liquid is added. Pure starches like potato, arrowroot, corn, and tapioca starches are best used in a ratio of 1 to 2 teaspoons per cup of liquid. These starches should be premixed with cold water to hydrate the starches and avoid clumps. Xanthan and guar gums are very effective thickeners, best used sparingly at a little less than ⅛ teaspoon per 1 cup of liquid; both are commonly used in gluten-free flour mixes to help mimic the action of gluten.

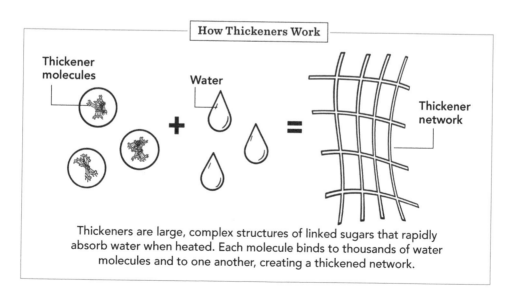

How Thickeners Work

Thickener molecules

Water

+

=

Thickener network

Thickeners are large, complex structures of linked sugars that rapidly absorb water when heated. Each molecule binds to thousands of water molecules and to one another, creating a thickened network.

WHY DO FRIED RICE RECIPES CALL FOR DAY-OLD RICE?

THE SCIENCE Rice is basically starch molecules that are packed away in tiny granules. During the process of cooking rice, the starch granules absorb water, causing them to swell, burst, and release their delicious starchy payload. When cooked rice is cooled down in the fridge, the starch molecules slowly recrystallize in a process known as retrogradation and transform into a form of starch called resistant starch. This crystalline starch behaves more like a fiber and reacts to frying differently than regular starch does. Researchers have found that when potato varieties rich in resistant starch were fried, they absorbed only 1 percent of their weight in cooking oil, compared with regular potato varieties, which absorbed 5 percent of their weight in cooking oil.

Even better, resistant starch has fewer calories. Normally, your body processes starches by releasing an enzyme known as amylase into your digestive system. Amylase chews up the starches and transforms them into sugars that can be easily absorbed by your small intestine. But the crystalline structure of resistant starch makes it difficult for amylase to break it down, so resistant starches bypass your small intestine and end up being digested by beneficial microbes in your colon that have special biochemical machinery to ferment them. Other cooked starchy foods, like potatoes, bread, and pasta, also go through the process of retrogradation during cooling and form resistant starches.

KITCHEN TAKEAWAY When a fried rice recipe calls for day-old rice, be sure to use it. Day-old rice contains a good amount of resistant starch, which fries better than freshly cooked rice. If you don't heed this advice and make a fresh pot of rice for your recipe, the fried rice will turn out soggy.

CHAPTER 2

Flavor Basics

FLAVOR IS WHAT MAKES FOOD ENJOYABLE. Without the complex chemistry behind flavor, food would be incredibly bland and boring. I would say everything would taste like cardboard, but even the sad flavor of cardboard is generated from a diverse array of compounds. The exciting part about flavor is that it's something you can control in the kitchen and harness to cook up delicious food. In this section, I'll answer some of the most common questions about the way different foods taste and explain how you can use that knowledge to infuse your dishes with even greater flavor.

How Do We Experience Taste?

THE SCIENCE The little bumps found on your tongue are called papillae and are lined with taste buds. Each taste bud contains 10 to 50 sensory cells that have protrusions called taste hairs. When a taste molecule, like sugar, encounters these taste hairs, specially shaped proteins that line the surfaces of the hair bind to the molecule and initiate a cascade of chemical signals inside the sensory cell. The sensory cell then releases neurotransmitters that interact with nerve fibers attached to the outside of the cell. The nerve fibers send chemical and electrical signals through a series of nerve bundles attached to one another like the wires in an electrical cable. These signals eventually make it to the region of the brain that perceives and processes them into taste sensations. All of this takes place in a few hundred milliseconds.

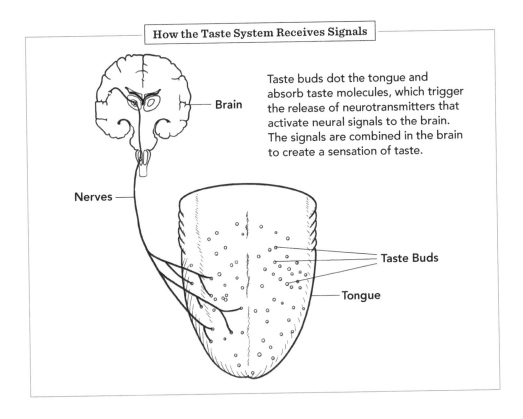

How the Taste System Receives Signals

Brain

Taste buds dot the tongue and absorb taste molecules, which trigger the release of neurotransmitters that activate neural signals to the brain. The signals are combined in the brain to create a sensation of taste.

Nerves

Taste Buds

Tongue

The human tongue contains 2,000 to 8,000 taste buds. Half of these taste buds contain sensory cells that interact with all five basic elements of taste—sweet, sour, bitter, salty, and umami. The other half specialize in specific taste elements and are involved with registering the intensity of what you are tasting. There is a long-held misconception that there are zones on the tongue for each taste bud—sweet taste receptors in the front, bitter receptors in the back. The truth is that each of the taste receptors are present in equal proportion across the tongue, except for bitter taste buds, which are, in fact, found in high proportion at the back of the tongue. Scientists believe that this region of the tongue is especially sensitive to bitter flavors so poisonous or rotten foods can be spit out before they are swallowed.

KITCHEN TAKEAWAY Trust your tongue. Millions of years of evolutionary trial and error went into the design of the human taste system. The tongue is one of the most sensitive sensors available and can easily detect a broad range of chemicals. If you taste something off or funny in your food, chances are there's probably a good reason for that sensation and you should spit it out.

Q WHAT FACTORS CAN AFFECT THE WAY WE EXPERIENCE FLAVOR?

THE SCIENCE The science behind flavor perception is a complex business. Much of the perception of flavor is experienced by the olfactory system of the nasal passages; only the five basic tastes and sensations (think minty coolness and spicy heat, for instance) are experienced by the tongue. When food is eaten, flavor molecules make their way to the back side of the tongue and eventually migrate to the back of the nasal cavity, where they are perceived as smell. The brain combines these experiences together to create a complete sensory picture of the flavor. This is why it's so difficult to taste anything when your nose is stuffed up from a cold or allergies; excess mucus creates a barrier between the smell receptors and the flavor molecules.

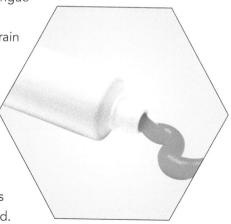

Flavor perception starts with the food itself, since the food's physical and chemical properties can change the way flavor molecules are released.

THE SCIENCE BEHIND FLAVOR PERCEPTION IS A
COMPLEX BUSINESS

The types of proteins, carbohydrates, fats, and salts found in a food can affect the rate at which flavor and aroma molecules are released from it. Temperature plays a role in flavor perception, too, because it affects the evaporation rate of each flavor molecule, which in turn affects how quickly a flavor molecule makes its way to your taste and smell receptors.

Once these flavor molecules make their way from the plate to your mouth, other factors come into play. The brain will integrate multiple sources of flavor and create different flavor experiences depending on the combinations. Consider how a glass of orange juice tastes when you drink it at breakfast versus right after brushing your teeth. The coating of bacteria on your tongue also affects flavor as it slowly breaks down last night's meal and releases potent-smelling compounds that are perceived by the olfactory receptors. The bacteria on your tongue also digest the food you're currently eating and produce different aroma compounds that are dependent on your unique tongue microbiome.

The olfactory system can also undergo sensory fatigue, which is a fancy way of saying your brain gets tired of sensing one flavor and just stops perceiving it all together. The brain responds to new stimuli much more readily than it does to old, tired ones. If you're eating a food that is heavy on one note, your perception of that flavor may weaken over the course of a meal. If you eat side dishes alongside your main meal, you may be less likely to experience this kind of sensory fatigue.

KITCHEN TAKEAWAY Mixing flavors during a meal is a good way to stimulate the olfactory senses and reduce or prevent sensory fatigue. When preparing a dish, consider how different combinations of strong flavors can interact, overwhelm, or play off one another and evolve over the course of a meal, in the same way that the scent of a perfume can change over time.

Are Taste and Flavor the Same Thing?

THE ANSWER No

THE SCIENCE If one person says that a meal is "flavorful" and another calls it "tasty," you can assume that both people share the same sentiments. In scientific circles, however, the concepts of taste and flavor are considered similar but different phenomena. Taste is just one aspect of flavor and refers to the physical sensations experienced when very specific classes of food compounds interact with the five basic taste buds on the tongue, eliciting sensations of sweet, sour, bitter, salty, and/or umami (savory). These basic tastes signal to the brain that there are sugars (sweet), acids (sour), poisons (bitter), minerals (salty), and/or amino acids (umami) in your food.

Flavor, on the other hand, refers to the kaleidoscope of subjective experiences that can occur when food is eaten and taste is integrated with smell, sound, color, texture, physical sensations, emotions, and memories. Flavor can be seen as the holistic experience of food, and it is highly dependent on both taste and aroma. Sense of smell is activated when aromatic compounds released by a food travel into the nostrils through sniffing or the back of the mouth along a passage called the nasopharynx. The process is known as retronasal olfaction. Signals for other sensations and textures, like spicy, hot, cold, crunchy, soft, and creamy, are also sent to the brain. Once the brain receives these signals, it processes and connects the current experience with older memories and emotions to try to find something familiar about the food you're eating. Much like how an intricate web of sensations, memories, and feelings can be concocted through a combination of chemical aromas in a perfume, the complex flavors of a well-prepared meal can bring out a dazzling array of subjective experiences.

KITCHEN TAKEAWAY Take the time to think about how different tastes, aromas, and textures elicit different responses in your mind. These sensations are an important aspect of how you experience food, and learning how you experience the nuances of food can help you enjoy each meal that much more.

Q CAN YOUR ENVIRONMENT AFFECT HOW FOOD TASTES?

THE ANSWER Yes

THE SCIENCE Have you ever wondered why airline food always tastes so bad? What may surprise you is that the bad taste you experience in an airplane has more to do with the environment you're in than the actual food itself (although I'm sure that plays a part in the experience, too). The combination of dry air, low cabin pressure, and loud engine noise suppresses the sensitivity of taste buds and the sense of smell. These environmental factors most strongly affect the experience of saltiness and sweetness, and airline food manufacturers add 30 percent more salt and sugar to the food served on planes to compensate for this effect. But, according to a 2010 study by the Fraunhofer Institute for Building Physics, while loud noises suppress sweetness, they enhance savory umami taste. This may be why tomato juice, which contains a high amount of umami-enhancing glutamates, is a common choice for airline passengers.

Music and sound also have incredible effects on the tastes and flavors of foods. Studies conducted by Charles Spence, a professor of experimental psychology at Oxford University, show a powerful relationship between sound and the experience of eating. Spence and other researchers have found that test subjects can consistently pair the basic tastes of sweet, sour, salty, bitter, and umami with certain tones and pitches. When the researchers played music with what one would characterize as "sweet" tones while the subjects ate toffee, the subjects reported a sweeter taste. When music with "bitter" tones was played (see the Kitchen Takeaway below), the subjects reported a bitter taste in the same toffee. In another experiment, Spence and his colleague designed music that amplified the creaminess and sweetness of chocolate, which is now being used by chocolatiers to improve the chocolate-eating experience for their customers. But sound-taste pairings aren't just for sweets. Many fine dining restaurants use this concept to enhance their ambiance and the experience of their patrons.

KITCHEN TAKEAWAY Experiment with different songs to see how they affect your taste buds. Playing the right melody with the right dish could really bring out the richness of a meal. Low-pitched sounds played on brass instruments tend to magnify bitter flavors, whereas high-pitched notes played on a piano are associated with sweetness.

IS THERE A SCIENTIFIC REASON WHY CERTAIN FOODS AND FLAVORS COMPLEMENT EACH OTHER?

THE ANSWER Yes

THE SCIENCE Some foods—peanut butter and jelly, bacon and eggs—just taste better together. And the same goes for certain food and drink pairings, like pinot gris with seafood and zinfandel with highly spiced food. To explain this phenomenon, English chef Heston Blumenthal developed the food pairing hypothesis, which suggests that ingredients and foods that share one or more flavor molecule will combine better on the palate than those that do not. This hypothesis has helped contemporary restaurants design unusual food pairings that taste delicious. For example, chocolate and blue cheese share at least 73 different flavor molecules and make for an interesting partnership. Other unexpected pairs that this work has uncovered include strawberries and peas, caviar and white chocolate, and harissa and dried apricots.

KITCHEN TAKEAWAY For most home cooks, the food pairing hypothesis is not much more than a fun parlor game, but it can be fun to experiment. An excellent resource is the website FoodPairing.com, which has catalogued thousands of flavor molecules in meats, vegetables, spices, juices, and other ingredients to help you develop new food pairs across all cuisines.

Is Umami Really a Thing?

THE ANSWER Yes

THE SCIENCE Japanese chemist Kikunae Ikeda first coined the term "umami" in 1908 to refer to a pleasant taste experienced when eating savory foods. Ikeda first encountered umami when he tasted an especially delicious broth prepared by his wife. Curious, he asked her what was in it. She explained that she had added a bit more kombu, a culinary seaweed, than usual. Ikeda became interested in determining the exact compound that produced this umami taste. After a year of painstaking extractions from several pounds of kombu, he discovered that the amino acid glutamic acid, or glutamate, was responsible for producing the umami sensation.

Glutamates can be found naturally in many different foods, including seaweed, soy sauce, miso, tomato sauce, aged cheeses, and yeast extract. These ingredients have been used for centuries to add a punch of flavor to savory dishes.

But umami taste is produced by more than just glutamates. Inosinates and guanylates, compounds formed from the breakdown of DNA by enzymes produced naturally by plants, animals, fungi, and bacteria, can also elicit the taste of umami. Foods rich in these compounds include fish sauce, anchovies, broth, mushrooms, and meat.

While the chemical principles behind umami were discovered in 1908, the concept of umami was met with resistance by the scientific community for many decades. However, in 1996, researchers first discovered the existence of a human glutamate taste receptor. The later discovery of several other amino acid taste receptors that responded to umami-associated compounds added credibility to the idea that umami was a fifth basic taste.

Umami taste perception is believed to have evolved to allow humans to detect proteins in foods, since glutamate is a key component of many proteins. Much like how sweetness helps us detect carbohydrates, saltiness helps us detect minerals, sourness helps us detect acids, and bitterness helps us detect toxins, the ability to taste umami may have been a way for early humans to quickly figure out which foods were rich in nutritionally important proteins and amino acids.

KITCHEN TAKEAWAY If you want to really beef up the flavor of a savory dish, add a trifecta of glutamates, inosinates, and guanylates. These compounds act synergistically to create an umami sensation of much greater intensity than each can achieve individually. Adding a few teaspoons of Worcestershire or fish sauce (which contain inosinates and guanylates) and soy sauce (which contains glutamates) will do wonders to deepen the savory flavors in a dish.

UMAMI TASTE PERCEPTION
IS BELIEVED TO HAVE EVOLVED TO ALLOW HUMANS TO
DETECT PROTEINS IN FOODS

WHY DOES SALT MAKE FOOD TASTE SO GOOD?

THE SCIENCE Paul Breslin and Gary Beauchamp were both researchers at the famed Monell Chemical Senses Center in Philadelphia in the mid-1990s. The two were puzzled by the ability of salt to enhance flavors in food, because all published research up to that point had shown salt to suppress or have little effect on flavors in isolation. In 1997, the researchers set out to uncover what exactly was going on between salt and our taste buds.

They subjected their study participants to different combinations of a bitter compound, a sweet compound, and/or salt. They found that when the subjects tasted the samples containing the bitter and sweet compounds together with salt, they reported a noticeable increase in the subjective experience of sweetness compared with the same sample without salt. On the other hand, when only the sweet compound and salt were present, there was no reported increase in the perception of sweetness. The researchers concluded that one way salt enhances flavor is by selectively suppressing bitterness, which allows other tastes and flavors to come to the forefront. Maybe that's why low-sodium foods are so poorly received by consumers. We need the salt to experience other flavors to their fullest.

KITCHEN TAKEAWAY Salt not only enhances savory flavors, it brings out sweetness by reducing the perception of bitterness. Test for yourself the effect of salt on flavors. Add a pinch of salt to foods and drinks that normally don't call for it (like fruits, ice cream, etc.), and see how it affects the taste. While many foods do not subjectively taste bitter, they still contain molecules that bind to our bitter receptors and integrate as a background signal to our brain that bitterness exists in the food. With just a little salt, those molecules can be suppressed, allowing extra sweetness to come through.

Q DOES THE KIND OF SALT I USE MATTER?

THE ANSWER Maybe

THE SCIENCE At the most fundamental level, all salts are chemically the same; they are all made up of sodium chloride. But the key differences between each kind of salt come down to the impurities present and the size, shape, and texture of the salt crystals. These distinctions produce subtle flavor effects when salts are added to food during cooking.

Table salt is mined directly from the earth and has very fine, crystalline grain structure. It is iodized (fortified with sodium iodide) to increase our dietary intake of iodine, a mineral important for normal thyroid function. It also contains anti-caking agents to prevent clumps in the salt shaker. The net result is that table salt tends to taste saltier compared with other types of salt because it more readily dissolves on the tongue and in food. The finer crystals also compact more tightly than coarser forms of salt, so more salt granules will fit in a teaspoon versus other types. Some people say they can taste the iodide, which presents as a bitter, metallic aftertaste.

Himalayan salt is also mined and is quite similar to table salt, except that it contains a bit of iron and copper, which gives it that pink hue and very minor flavor differences. This salt is generally produced in a coarser grain than table salt.

Kosher salt is a coarse-grain salt that's usually mined from salt deposits. Diamond Crystal brand kosher salt is produced using the patented Alberger process, a

THE LOWER DENSITY OF KOSHER SALT RESULTS IN A LESS INTENSE LEVEL OF PERCEIVED SALTINESS

mechanical evaporation and steaming method that produces a low-density flaked salt with high solubility that is coarser than table salt. The lower density of kosher salt results in a less intense level of perceived saltiness, and the greater crystal size means less salt per teaspoon than the same measure of table salt. Kosher salt is also free of sodium iodide, which results in a cleaner flavor than that of iodized table salt. Because kosher salt has such unique flavoring properties, replacing it with another type of salt can affect a recipe.

Then there's sea salt. Because it's produced by evaporating seawater, sea salt contains a variety of other minerals like magnesium, calcium, bromide, and naturally occurring iodide. These components can vary depending on the composition of the seawater the salt came from, but most people can't tell the difference between the flavors in sea salt versus other salts. Different evaporation methods can lead to different levels of coarseness, from a flaky, hollow, quick-dissolving salt to extra-coarse granules that struggle to adhere to food.

KITCHEN TAKEAWAY Different types of salts can produce different results in food, mainly due to the coarseness of the crystals and their solubility in water. Kosher salt is a better choice if you want a milder salty flavor, whereas table salt is good for situations where you want salt to dissolve quickly, like in baking recipes. The type of salt does matter when you are following a recipe. If a specific measure of salt is called for, make sure you know what kind of salt the author of the recipe recommends, because using the wrong type can result in a finished dish that is either too salty or not salty enough.

What Makes Fat So Tasty?

THE SCIENCE Scientists have found that mammals possess a taste receptor, known as CD36, that binds to fat molecules. In a study involving mice genetically engineered without that gene, researchers discovered that the altered mice did not have much of a taste for fat, while the unaltered mice continued to gobble it up. Another study found that humans who express higher levels of the CD36 receptor were much more sensitive to the taste and smell of fat in food than were those who expressed lower levels of the receptor. The CD36 receptor is also involved in a neural circuit that releases serotonin, a neurotransmitter involved in happiness and well-being, when exposed to fats.

However, our affinity toward fatty foods isn't all about genetics—the physical and chemical properties of fats also improve our experience of other flavors and textures. Fat acts as a solvent for certain fat-soluble flavors like those found in woody herbs, spices, and meat. Fat can also slow down the release rate of flavor molecules, prolonging their exposure to our olfactory system and increasing our enjoyment. Plus, frying or roasting with fats creates very desirable crispy textures and flavor molecules through the Maillard reaction (see page 12) that can't be replicated with other cooking methods.

KITCHEN TAKEAWAY The secret to most great sauces is fat! For an easy, smooth sauce, deglaze a pan by adding wine or stock after searing steak or lamb chops, and use a wooden spoon to scrape up the delicious browned bits at the bottom of the pan for flavor. Then, swirl in several tablespoons of butter, one at a time, to create a creamy emulsion.

Q WHY DOES CILANTRO TASTE SOAPY TO SOME PEOPLE AND NOT TO OTHERS?

THE SCIENCE There are about half a dozen compounds responsible for cilantro's characteristic flavor. Most of them are grouped under the class of chemicals known as aldehydes, and there is evidence that some individuals have a genetic predisposition to be highly sensitive to aldehydes, experiencing them as a soapy flavor. This variant of the olfactory receptor gene occurs in nearly 14 to 21 percent of East Asians, Africans, and Caucasians, versus 3 to 7 percent of South Asians, Hispanics, and Middle Easterners. Interestingly, aldehydes also happen to be a by-product of making lye soap, which may be why some people's sensory experiences draw the connection between cilantro and soap.

KITCHEN TAKEAWAY If you really can't stand the taste of cilantro, here's a trick you can try. Cilantro contains enzymes that gradually break down those soapy aldehydes into compounds with little to no aroma. Chopping, mincing, or pureeing the leaves into a paste will release the enzymes; let the cilantro sit for 15 minutes, then use.

Is Terroir Really a Thing in Food and Wine?

ANSWER Not really

THE SCIENCE Terroir is vaguely defined as the set of environmental factors that affect the characteristic taste and aroma of a food produced in a specific region. The term is most often used to describe wine, but it has also been studied in terms of chocolate, coffee, hops, peppers, and tomatoes. The environmental factors included in terroir are soil pH and mineral composition, climate, and farming practices, and they are believed to affect the flavor characteristics of a crop and thus the food and wine made from it. For instance, you may read about terroir on a wine label, where a wine's minerality or earthiness is ascribed to the geography of where the grapes used to make it were grown.

This all sounds great, but there is quite a bit of controversy over the concept of terroir. Mark Matthews, a viticulture professor at the University of California, Davis, believes terroir is a marketing ploy, with little basis in scientific fact. There have been scientific papers published that show soil cannot directly influence the taste of a wine or food because plants are unable to absorb the complex minerals that make up the soil they grow in. That said, environmental factors like soil certainly do play a role in affecting the flavors of a food through their indirect impact on water retention, nutrient composition, and plant physiology. But terroir seems to be a catch-all term tailored to the marketing needs of the wine and food industry, rather than a precise set of scientific factors that have a real impact on the character of a food.

TERROIR SEEMS TO BE
A CATCH-ALL TERM MORE SUITED TO MARKETING NEEDS

CAN A MARINADE INFUSE A FOOD WITH FLAVOR?

THE ANSWER It depends

THE SCIENCE Marinades typically include an acidic ingredient like vinegar, wine, or yogurt mixed with other flavoring ingredients. The theory is that the acids in the marinade break down the tissue on the surface of the food, allowing the marinade to be absorbed and flavor the food. The problem is that this theory runs counter to reality. In tests done by America's Test Kitchen, marinade flavorings penetrated only 0.03 to 0.1 inch/1 to 3mm in chicken breasts marinated for 18 hours. The common prescription for this problem is to allow the marinade to sit with the food for longer, which doesn't really help in an appreciable way and can lead to another issue. Food, especially seafood, that has been left to marinate for too long can become mushy as the acid dissolves the proteins. The only marinades that appear to truly infuse foods with flavor are salt-based ones and those that include generous amounts of soy sauce or fish sauce. That's because salt breaks open muscle cells, which helps the marinade penetrate deeper into the tissue. Higher concentrations of salt cause more cell disruption and improve marinade penetration, but be careful not to make your marinade too salty—at a certain point that saltiness can affect the flavor of the meat in a negative way.

KITCHEN TAKE-AWAY For the best flavor, use a marinade that has a high concentration of salt. The easiest way to do this is to include salt-rich ingredients like soy sauce or fish sauce. Mix ¼ cup or ½ cup of either (or a combination) with an equal volume of other liquid ingredients like vinegar, or use them alone. It can take about two hours for meat to soak up the flavor of a salt-based marinade.

What Gives Black Pepper Its Punch?

THE SCIENCE Piperine is the main flavor molecule found in black pepper, alongside several other minor molecules that add citrusy, woodsy, and floral notes. Though piperine activates the same receptor as the spice molecule in chiles—capsaicin—it is only about 1 percent as spicy as capsaicin, so the bite of black pepper fades much faster than that of a very hot pepper.

Our enjoyment of pepper goes beyond just the flavor it adds to food. Piperine has been found to stimulate salivary glands and bile secretions, both of which aid in digestion. It also significantly inhibits the activity of certain liver enzymes involved in detoxification, which allows health-promoting compounds found in herbs and spices to remain in the bloodstream longer, improving their availability and rate of absorption into our bodies. It's possible that we've developed a psychological association between health and the consumption of pepper, which adds to our pleasure of eating peppery foods.

KITCHEN TAKEAWAY Black pepper loses its potency when exposed to light, which converts piperine into the tasteless isochavicine. Black pepper also loses its flavor through evaporation, especially if pre-ground. For stronger flavor, it's best to grind pepper fresh right before you use it and store the whole peppercorns in an airtight container away from sunlight and heat. Pepper should be added near the end of cooking or right before serving to retain its flavor.

 ## WHAT GIVES GINGER ITS HEAT AND SPICINESS?

THE SCIENCE The shape and structure of a flavor molecule are important in the perception of flavor and taste. Gingerol is the molecule responsible for the flavor and pungency of fresh ginger, and its chemical structure is similar to that of capsaicin, the molecule behind the spiciness in chiles. Both molecules interact with the receptors on the lookout for spicy molecules, called capsaicin receptors, which then transmit a signal to the nervous system that causes the sensation of heat and pain. Any molecule that can strongly bind to these capsaicin receptors will cause this effect, including the pungent molecules found in wasabi, horseradish, and black pepper. But it's clear that the effects of chewing ginger are quite different from biting into a habanero, which is due to slight differences between the molecular structure of gingerol and capsaicin. The gingerol molecule is slightly shorter in length and contains an oxygen atom that juts out of the molecule like a tiny appendage, whereas capsaicin contains a nitrogen atom that's embedded inside its linear molecular shape. As a result, capsaicin has a better fit and binds more strongly to the capsaicin receptor than gingerol does.

KITCHEN TAKEAWAY If you want that extra kick, be sure to use fresh ginger. Freshly grated ginger is more pungent and flavorful than ground ginger because gingerol, which slowly decomposes over time, is most present in fresh ginger.

Why is Saffron So Highly Prized?

THE SCIENCE The wispy threads of saffron you buy in little tubes are actually the stigmas from the saffron crocus flower *(Crocus sativus)*; it takes about 13,500 stigmas hand-picked from 4,500 flowers to produce one ounce of saffron threads.

The molecule responsible for saffron's unique flavor, safranal, was discovered in 1933 by Richard Kuhn and Alfred Winterstein. Saffron contains a molecule called picrocrocin that is broken apart by an enzyme during its harvesting and drying process, releasing safranal. This compound makes up about 70 percent of the total volatile oil mass of saffron. Researchers later discovered that another compound, lanierone, also contributes to the coveted flavor of saffron, despite making up only a small fraction of its essential oil. Together, these flavor compounds impart a subtle, hay-like flavor and aroma to saffron that's sometimes described as floral, honey-like, bitter, or pungent. Saffron is used in Indian and Middle Eastern cooking its golden hue, which adds vibrant color to dishes.

KITCHEN TAKEAWAY Because of saffron's very high price, quite a bit of what is sold is counterfeit or adulterated. To determine if yours is the real thing, soak a few threads in hot water for 5 to 20 minutes. Real saffron remains intact and will turn the water an even, uniform color; fake saffron will disintegrate and quickly bleed out artificial dye. Saffron oils are sensitive to heat, air, and light, so it's best to store it in an airtight container in a cool, dark place.

Q WHAT GIVES NUTRITIONAL YEAST ITS DISTINCTIVE FLAVOR?

THE SCIENCE Nutritional yeast is a pure strain of *Saccharomyces cerevisiae*, the same yeast used in beer brewing and baking. The yeast is first cultured on a cheap source of sugar and nutrients (e.g., molasses, beet sugar, etc.) for several days. Once the yeast has been cultured, it is heated to pasteurization temperatures to deactivate it, then kept at a temperature that allows enzymes native to the yeast to break apart the cell walls and release the compounds inside. Protease enzymes chew up the yeast proteins and produce the amino acid glutamic acid, which enhances umami flavors. Nuclease and phosphatase enzymes break down its DNA into smaller nucleotide subunits like inosinate and guanylate. Together, glutamic acid, inosinate, and guanylate produce a synergistic effect that results in a very strong umami flavor.

Nutritional yeast also contains a high concentration of a peptide (a small chain of amino acids) known as glutathione, which is believed to elicit what is now known as kokumi, a taste first identified in 1989 by Japanese researchers at Ajinomoto, a Japanese flavor company. Kokumi remains a controversial taste, and many Western taste researchers remain skeptical about its contribution to flavor. Kokumi does not evoke an actual subjective taste; rather, it heightens other tastes and flavors, including umami, and lengthens the time that they're experienced. If taste were experienced as music, kokumi dials up the volume and adds an encore. Nutritional yeast's combination of glutathione and the three umami compounds (glutamic acid, inosinate, and guanylate) make it a very potent flavor enhancer.

KITCHEN TAKEAWAY Nutritional yeast is an excellent vegan substitute for animal-derived flavorings like aged cheese, fish sauce, meat extract, and shrimp pastes. It has a natural nutty, cheese-like flavor that complements savory foods and can be used in the same way as grated cheese (try it sprinkled over popcorn) or to add depth of flavor to soups, stews, and gravies as you would use soy, Worcestershire, or fish sauce. It's also a great option to punch up the flavor in a dish without resorting to salt.

Is It Better to Use Herbs Fresh or Dried?

THE SCIENCE The flavorful, pungent essential oils that attract us to fresh herbs are actually compounds created by the plants as a defense against pests. Many of these compounds are quite volatile, meaning that they instantly evaporate once the plant tissues are broken or crushed to quickly deter insects or animals with their strong aromas. During the process of drying herbs, the more volatile flavors are lost and only the heavier oils remain. Also, depending on the drying method and temperature used, the herbs' volatile oils can undergo decomposition and oxidation, not only muting some of the original intensity of the herb, but in some cases, creating bitter flavors. Delicate, leafy herbs such as chives, tarragon, and parsley can lose much of their flavor when dried.

However, there's a place for dried herbs in the kitchen, especially since these flavor enhancers can be stored for long periods. When making a soup, sauce, or stew that will be cooked or simmered for longer than 10 minutes, it's better to use dried herbs, since the heavier oils that remain in dried herbs are much less affected by heat and evaporation than the volatile oils in fresh herbs. Also, some tougher woody herbs, like rosemary, oregano, thyme, and sage, retain their flavors even when dried.

KITCHEN TAKEAWAY To preserve the oils and bright taste of fresh herbs, serve them raw in salads, as garnishes on top of finished dishes, or as last-minute flavor enhancers stirred into a dish right before serving. Dried herbs are better in dishes that will be cooked for some time, like chilis, stews, soups, and long-simmering sauces. The rule for substituting dried herbs for fresh is to use one-third of the amount, since dried herbs contain less water and are therefore more concentrated than fresh herbs.

> ## DURING THE PROCESS OF DRYING HERBS, THE MORE VOLATILE FLAVORS ARE LOST

WHEN SHOULD I USE LEMON ZEST VERSUS LEMON JUICE?

THE SCIENCE The components of lemon juice are water, sugar, and acid, and the primary acids in lemon juice are ascorbic acid, better known as vitamin C, and citric acid. These acids are wonderful for adding sharp, bright flavors to a dish or changing the pH in positive ways—for example, the acids in lemon juice help remove fishy odors (see page 69). Lemon juice is also useful for reducing browning in avocados and apples (see page 95).

Lemon peel (in fact, all citrus peels) contains no acid and adds lemon flavor through the flavor compounds limonene, citral, and terpinine present in its essential oil, which is released when the peel is grated to make zest. Lemon zest contains very little water, compared with lemon juice, so it's great for dishes where you want lemon flavor but not much liquid.

And, because zest contains no acid, it adds flavor without tartness. Lemon zest is also a good choice if you want to add lemon flavor to a baked good because it allows you to do it without changing the pH or water content of the batter. If you use juice instead, without making other adjustments, the additional acid and liquid can throw off the chemistry of the batter, affecting the leavening and texture of the finished product.

KITCHEN TAKEAWAY Use lemon juice in preparations that need the brightness of acidity and can take the extra liquid, such as soups and marinades, and squeeze it over fish (to reduce fishiness) and avocado (to prevent browning). Use zest when you want lemon flavor without the acid or added liquid, such as in dishes with dairy (where acids could cause curdling), simple syrups, and baked goods. If you can, add zest at the last minute to avoid evaporating the volatile essential oil. Lemon zest is also a good choice when you are cooking with aluminum or copper cookware, since acids can leach those metals.

CHAPTER 3

Meat, Poultry, and Fish

ANIMAL MEAT WAS ONE OF THE FIRST RAW FOODS TO BE THROWN ON THE FIRE AND ROASTED INTO A SUMPTUOUS MEAL BY EARLY HUMANS. But how exactly are these hunks of protein, water, and fat transformed into delicious, succulent entrées? As it turns out, the mystery of this transformation is all in how heat, moisture, and time take the essential components of meat and recombine them into a mouthwatering arrangement of flavors, colors, and textures.

Why Are Meats Different Colors?

THE SCIENCE Dark, or red, meat comes from the muscles in an animal that use a lot of oxygen to support movement, such as the wings, legs, and thighs. These muscles contain an abundance of myoglobin, an iron-rich protein that helps mammalian tissues transport and store oxygen from the blood system, and it is myoglobin that gives dark meat its purplish-reddish hue. When myoglobin is heated to 170°F/77°C or higher, it transforms into metmyoglobin, which gives the interior of well-done meat its brownish gray color. The chemical conversion of myoglobin to metmyoglobin is also the reason why freshly cut meat will turn from a bright red or pink color to a dark gray as the meat becomes older. Because consumers believe something has gone wrong with the meat if it's turned gray, meat manufacturers will add carbon monoxide to packaging, which binds to myoglobin to form a red pigment and maintain the meat's red color. Pork, which tends to be lighter in color than beef, has a lower concentration of myoglobin because the pigs used to produce pork are younger and smaller, so their muscles are less developed.

What we think of as white meat in chicken or turkey comes from the parts of animals that go through shorter bursts of activity and don't rely on as much oxygen.

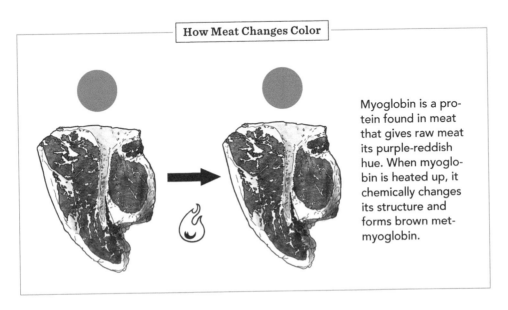

How Meat Changes Color

Myoglobin is a protein found in meat that gives raw meat its purple-reddish hue. When myoglobin is heated up, it chemically changes its structure and forms brown metmyoglobin.

As a result, there is less need for myoglobin in these meats and these meats stay white. Some poultry, like ducks, spend a lot of their time flying and using oxygen, which is why duck meat is nearly entirely dark colored.

KITCHEN TAKEAWAY When it comes to beef or lamb, it's normal for the meat to turn from red to gray, and that change in color doesn't necessarily mean that the meat has gone bad. Instead, check to see if the meat has an off aroma or feels slimy to the touch—that's how you know the meat is no longer safe to eat.

WHAT IS THE DIFFERENCE BETWEEN TENDER AND TOUGH CUTS OF MEAT AND HOW THEY COOK?

THE SCIENCE Most of what we think of as meat is skeletal muscle tissue, which is made up of protein filaments that are designed to contract against one another and produce motion. These filaments are bundled together to form thin, elongated muscle fibers. Whether a cut of meat is tender or tough is dependent on the density of protein filaments and the amount of collagen they contain. Collagen is composed of connective proteins, whose job is to keep the muscles together. Meat that comes from the muscles that are regularly exercised, like those in the shoulders and legs, is tougher because the filaments multiply in response to the muscles' continuous resistance and movement—and the more filaments there are, the thicker (denser) the muscle fibers are and the tougher the meat will be. Cuts of meat that come from regularly exercised parts of animals (like the chuck roast, pork butt, and brisket) also contain more connective tissue and collagen, which can give tough meat an unpleasantly chewy texture.

That being said, "tough" is a bit of a misnomer when it comes to meat. Certain cuts of meat are only tough if they are not cooked properly. To break down the muscle fibers, connective tissues, and collagen proteins, roasts and other cuts of meat with dense muscle fibers and high amounts of collagen require a moist environment and a longer cooking time—collagen melts and forms gelatin when heated in the presence of moisture for long periods. In other words, these cuts are tailor made for cooking methods like braising, simmering, and stewing.

The presence of water (or any liquid) helps to speed up tenderization of meat by hydrating and reacting with proteins like collagen through the process of hydrolysis; this is what occurs during braising and simmering. This tenderizing moisture can also be supplied by the meat itself through its own meat juices, as in the case of the low and slow smoking of large cuts like brisket or pork butt. The key here is to keep the temperature low enough that the meat maintains its moisture as the collagen begins to soften and then finally melts, transforming the meat from chewy to incredibly tender as the muscle fibers basically fall part after hours of cooking.

Finally, tough meat can also be tenderized by using a mallet to physically break the protein and muscle fibers in the meat before cooking.

Tender cuts, such as tenderloin, come from the relatively underutilized back and loin muscles of an animal and therefore have a lower muscle fiber density. Some tender cuts also have especially high proportions of fat, as they came from parts of the animal that stored fat while it was alive. Cooking methods for these cuts focus more on further developing flavor, in particular the tasty browning that results from the Maillard reaction through such dry heat techniques as searing, roasting, and grilling (for more on this, see Why Do Foods Brown? The Maillard Reaction, page 12).

KITCHEN TAKEAWAY All cuts of meat can yield tender results if matched to the correct cooking method. Long, slow cooking methods at lower temperatures give chewier cuts the time to break down into delicious tenderness; quick, high-temperature cooking methods can add extra layers of flavor to cuts that are already tender.

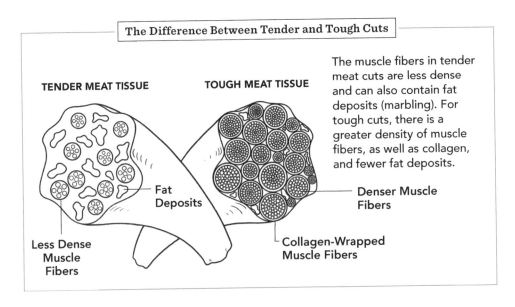

The Difference Between Tender and Tough Cuts

TENDER MEAT TISSUE

TOUGH MEAT TISSUE

The muscle fibers in tender meat cuts are less dense and can also contain fat deposits (marbling). For tough cuts, there is a greater density of muscle fibers, as well as collagen, and fewer fat deposits.

Fat Deposits

Denser Muscle Fibers

Less Dense Muscle Fibers

Collagen-Wrapped Muscle Fibers

Does Aging Meat Really Make It More Tender?

THE ANSWER Yes

THE SCIENCE Dry aging meat is the process of exposing it to carefully controlled temperatures and humidity levels, which will slowly alter its flavor and tenderness over time. Here's what happens during dry aging: Enzymes found inside the meat tissue, known as proteases, slowly break down the proteins that make up the muscle and connective tissues into amino acids and peptides (protein fragments) and increase the tenderness and enhance the flavor of the meat. Bacteria and mold will also grow on the surface of the meat during the dry aging process, much like they do on aged cheeses, and release enzymes that further break down the proteins in the meat. Dry aging will cause the meat to slowly dry out, resulting in a mummified exterior layer; for that reason, dry aging is usually reserved for very large pieces of meat, which can be trimmed of this dried layer after aging to reveal the juicy, tender interior.

KITCHEN TAKEAWAY The most-aged beef isn't necessarily the tastiest. Most of the tenderizing effects happen in the first 10 to 14 days of aging. And while beef can be aged well beyond the 30-day mark, improperly aged beef can start to rot and give off unpleasant odors.

WHY DOES MARBLING MAKE FOR A BETTER TASTING STEAK?

THE SCIENCE Marbling refers to the white flecks of fat interspersed among the muscle fibers in a piece of steak. It occurs when cows are fed a grain-based diet and the excess fat is stored in the areas of their bodies that get very little exercise. Some tender cuts of meat, like tenderloin, naturally have little to no marbling because they come from areas of the animal that do not store fat. The USDA grading system rewards marbling, with Prime designated as the highest marbling content.

Why the fuss over marbling? Most flavor molecules are fat soluble and concentrate in the fatty portions of an animal. Additionally, fat helps to trap moisture as a steak cooks, maintaining its juiciness. The fat also serves as a lubricant between

MOST FLAVOR MOLECULES ARE FAT SOLUBLE

AND CONCENTRATE IN THE FATTY PORTIONS OF AN ANIMAL

muscle fibers, which gives steak a silky texture and makes it easier to chew. Without marbling, we wouldn't have the savory flavors and fine texture we associate with a great-tasting steak.

KITCHEN TAKEAWAY The simplest way to find a well-marbled steak is to look for the USDA shield that signifies USDA Prime certification. The USDA Prime designation refers to the youngest beef, which is the most tender, with the most marbling. USDA Choice cuts are also high in quality but have less marbling than Prime cuts do. Beef cuts with the most marbling tend to come from the top of the cow—rib-eye, strip loin, and top sirloin steaks all have significant marbling.

Can a Marinade Tenderize Meat?

THE ANSWER It depends on the marinade

THE SCIENCE Marinades may add flavor to meat, but they don't necessarily tenderize it. In fact, acid-rich marinades (marinades containing vinegar, citrus, or wine) can toughen a meat by lowering its pH. All proteins have a pH at which they will have a zero net charge; this is known as the isoelectric point. The isoelectric point is special because proteins lose their ability to absorb water at this pH, which causes them to shrink and become tough. The approximate isoelectric point for most muscle proteins is in the range of 5.0 to 5.6. The pH of wine is 3.3 to 3.6, and vinegar and lemon juice have pHs of 2 to 3. So, you can see how the pH of a marinade can easily reach the isoelectric point of the animal protein, leading to tougher meat if the meat is left in the marinade for too long.

Acidic marinades can penetrate fish deeper and faster than they can meat or poultry because fish muscles are arranged in flaky, segmented sheets connected by a thin layer of connective tissue, whereas land animals have muscles composed of fibers held in tight, dense bundles that resist marinade penetration. Fish also contain much less collagen than land animals do, and the collagen they do have breaks down easily when exposed to acidic marinades. (See Does Citrus Juice Really Cook Seafood in Ceviche?, page 67, to learn more about the muscle fibers of fish.)

If you want to tenderize meat using a marinade, skip the acidic elements and use an enzyme instead. Tenderizing enzymes break proteins into shorter fragments; in the case of meat proteins, the enzymes slowly break apart the muscle fibers. The primary enzymes used for tenderizing animal protein are bromelain, papain, and actinidain. Bromelain is found in the fruit and stem of pineapple, and

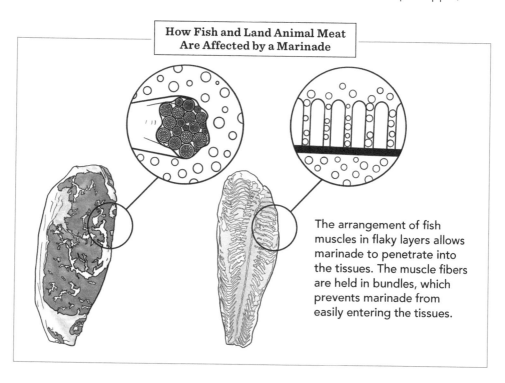

How Fish and Land Animal Meat Are Affected by a Marinade

The arrangement of fish muscles in flaky layers allows marinade to penetrate into the tissues. The muscle fibers are held in bundles, which prevents marinade from easily entering the tissues.

it works best in its fresh form because the enzyme can be deactivated by high temperatures; processed pineapple products like pineapple juice and canned pineapples do not contain active bromelain enzymes. Papain is found in papaya, and, like bromelain, will deactivate at higher temperatures. Actinidain is found in kiwifruit, though it doesn't have quite the protein busting powers of the other two.

KITCHEN TAKEAWAY Acidic marinades should be avoided if the goal is to tenderize meat, unless you are using the marinade for fish, which absorb acidic ingredients well because of their muscle structure. Acidic marinades will cause most meats to become tough. Pineapple, papaya, and kiwi are good sources of enzymes that can be used to tenderize meat. For the best tenderizing effect, the flesh of the fruit should be pureed and used as is or mixed with other marinade ingredients. Marinate tender meats in this mixture for at least 20 minutes and no more than 1 to 2 hours in the refrigerator to keep them from getting mushy; marinate tougher cuts in the refrigerator for up to 12 hours.

ACIDIC MARINADES CAN PENETRATE FISH DEEPER AND FASTER
THAN MEAT OR POULTRY

DOES BRINING MAKE MEATS JUICY?

THE ANSWER Yes

THE SCIENCE Normally, proteins in the muscle fibers of meat are hydrated by their own natural moisture content. Water molecules surround the proteins and bind to open pockets in the protein molecular structure. When meat is heated, the muscle proteins unravel and clump together through a process called denaturation. This rearrangement causes the protein fibers to shrink, tighten, and release water as meat juices. Brining can both increase the amount of liquid in the meat before cooking and reduce the amount of moisture lost/released during cooking. Brine, which contains a high concentration of salt, can readily enter muscle tissues and introduce more water. The salt in the brine denatures the proteins in a similar manner as the heat from cooking but does so without causing water to be released—in fact, just the opposite occurs. Proteins contain charged segments that interact with the salt ions, allowing the proteins to hold onto water better. As the proteins unravel, water molecules enter the deep molecular pores of the protein and bind to the cracks and crevices. The brine water becomes trapped there during the cooking process, resulting in a juicier meat.

KITCHEN TAKEAWAY Brining does help keep meats juicy. Use about 1 cup of table salt or 2 cups of kosher salt for every gallon of water. Most meats can be brined from 1 to 4 hours, depending on their size. The meat should be kept entirely submerged while it brines.

Should I Bring Meat to Room Temperature Before Cooking It?

THE ANSWER No

THE SCIENCE You see it in recipes all the time, the instruction to take a piece of meat (particularly a steak or roast) out of the refrigerator 30 to 60 minutes before you intend to cook it so it can come to room temperature. The thinking behind this, apparently, is that the meat will then take less time to cook.

The reality is that it takes quite a while to bring meat to room temperature. Tests that J. Kenji López-Alt, author of *The Food Lab*, did with steak showed that after 20 minutes at room temperature, the internal temperature of the steak had increased by less than 2°F. The rate that heat transfers to a steak from a hot pan is much higher than the heat transfer rate between room temperature air and steak, so it makes little sense to spend time leaving a steak or roast out before cooking it.

KITCHEN TAKEAWAY You will taste little difference between a steak that was left to warm at room temperature for 20 to 30 minutes before cooking and that went straight from the refrigerator to the pan, so skip the hype and leave your meat in the fridge until you're ready to cook it.

DOES SEARING A STEAK SEAL IN ITS JUICES?

THE ANSWER No

THE SCIENCE According to Harold McGee, author of the groundbreaking best-seller *On Food and Cooking*, the idea that searing a steak can seal in its juices is one of the biggest myths in cooking. He showed in his own kitchen experiments that, while searing a steak forms a crust, that crust is not waterproof—leave the cooked steak on a plate, and juice will leak from it. Despite McGee's observations, this myth persists, requiring other food writers and scientists to continue to debunk it with their own experiments. "Meathead" Goldwyn, barbecue expert and author of *Meathead: The Science of Great Barbecue and Grilling*, tested this by taking two steaks and searing only one of them before cooking. He then heated both steaks to the same temperature and weighed them afterward. Both steaks weighed the same, meaning that they contained the same amount of moisture, thus proving that searing the one steak did little to retain extra liquid.

One thing a quick sear does do, however, is create tasty flavor. When meat comes into contact with the high temperature of a hot pan, the Maillard reaction occurs, transforming amino acids into hundreds of delicious flavor compounds (for more details on this, see Why Do Foods Brown? The Maillard Reaction, page 12). That's why some roast recipes suggest that you brown the meat in a hot pan before slow cooking.

All that being said, it is important to note that, although a quick sear before cooking does not lock in a meat's juices, the method of cooking food completely over high heat for a short amount of time does result in less moisture loss than longer, lower-heat cooking methods produce, since a shorter cook time means less time for water to evaporate from the meat.

KITCHEN TAKEAWAY Searing produces intense flavor and color and minimizes moisture loss, compared with other cooking methods, if done quickly. To quickly sear a steak, use a dry (ungreased) stainless steel pan heated to 500°F/260°C or higher; the very high temperature will keep the steak from sticking to the pan (see Why Does Meat Stick to a Hot Pan?, page 60) while it cooks.

Does It Matter Whether You Cook Bacon in a Cold or Hot Pan?

THE ANSWER Yes

THE SCIENCE Bacon contains quite a bit of fat. One 8-gram/~0.3-ounce piece of bacon can contain 3.3 grams of fat, over 40 percent of its mass. Cooking bacon requires an understanding of how fat is affected by heat. If you place a cold strip of bacon in a hot pan, the meaty portion will brown before the fat has a chance to render and you'll end up with bacon that has a gummy, soft texture. If cold strips of bacon are placed in a cold pan and slowly heated up, the fat will liquify as the meaty portion browns and the rendered fat will aid in the crisping process.

KITCHEN TAKEAWAY For evenly browned, crispy bacon, starting in a cold pan over low to medium heat is best. Another excellent method is to bake bacon in the oven at 425°F/218°C; it'll take about 20 minutes for the bacon to crisp up perfectly.

WHY DOES MEAT STICK TO A HOT PAN?

THE SCIENCE You know the scenario—you heat up a pan, brown chicken cutlets on one side, and when you go to turn them, they are stuck like glue to the pan, to the point that the meat will tear if you try to yank them loose.

The reason for this problem is that the proteins in meat contain cysteine, an amino acid with one sulfur atom attached to it. Sulfur is very reactive and can form relatively stable chemical bonds. When meat hits a heated pan, the proteins unravel and expose their cysteines to the metal. The sulfur atoms react to the metal of the pan, forming a strong metal-sulfur bond that attaches the meat to the pan. Eventually, as the meat cooks on the pan, its surface heat causes the cysteines to break down, also breaking their bond with the pan.

KITCHEN TAKEAWAY When browning meat, have patience; once the cysteine breakdown point is reached, the meat will release from the pan on its own. Or, avoid sticking altogether by starting with a super-hot pan (475°F/246°C or higher)—the cysteines in the meat will break down pretty much on contact.

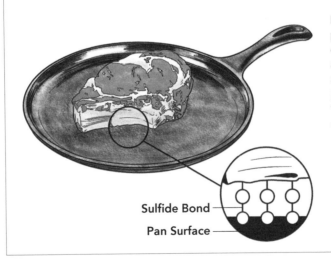

Why Meat Sticks to a Hot Pan

Meats and other protein-rich foods contain cysteine, a sulfur-containing amino acid. When the proteins heat up on the pan, the sulfur atom in the cysteine reacts to the metal in the pan to create a very strong sulfide bond that causes the meat to stick.

Sulfide Bond
Pan Surface

DOES BASTING MAKE MEAT MOIST?

THE ANSWER Not really

THE SCIENCE The team at *Cook's Illustrated* conducted an experiment to test the effects of basting a turkey breast every 20 minutes side by side with a control turkey breast that wasn't basted. A third turkey breast was included that wasn't basted but the oven door was opened every time the experimental bird was basted. They found that the moisture loss in all three was nearly the same. However, the basted turkey did show more browning than the other two, most likely because amino acids in the drippings triggered Maillard browning.

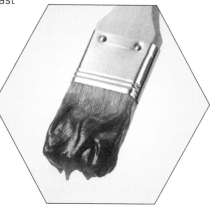

KITCHEN TAKEAWAY If a moist result is your objective, try brining instead of basting (see Does Brining Make Meats Juicy?, page 57).

Why Does Meat Dry Out?

THE SCIENCE First, let me challenge the use of "dry out." Juiciness and dryness are perceived subjectively by humans, and multiple factors affect these qualities. The amount of free versus bound water in a meat is one such factor. If the proteins in the meat bind strongly to the water molecules, there isn't a whole lot of water leftover for your mouth to experience juiciness. This is the case with beef jerky; it has a water content of 25 percent, yet it is experienced as dry. Other components like salt and mineral content can influence the way water is arranged and bound in the meat protein matrix.

The amount and types of fat in meat are also important factors in the perception of dryness. When meat fat melts during cooking, we perceive the slickness and fluidity of this fat as a kind of juiciness. Fat can also help trap water in the meat during cooking, preventing the meat from being perceived as dry. When

THE AMOUNT AND TYPE OF FAT
IN MEAT ARE ALSO IMPORTANT FACTORS IN THE PERCEPTION OF DRYNESS

meat is cooked, its proteins denature and shrink, causing some of its water content to be expelled in the form of what we would think of a meat juice. Over-cooking causes the proteins to shrink even further, expelling more water and causing the meat to feel dry. However, the fat in marbled cuts of meat can compensate for the loss of moisture by serving as a lubricant between the protein fibers. Some slow-cooked cuts of meat with high collagen content can remain moist because the collagen breaks down into gelatin, which contributes to a sense of juiciness. This collagen breakdown is the secret behind the deliciousness of a beef brisket smoked low and slow for many hours.

KITCHEN TAKEAWAY Each cut of meat should be cooked using methods that best fit its distinct composition of fat, water, collagen, connective tissues, and muscle fiber proteins. Understanding your meat and how it behaves relative to moisture, temperature, and time can help prevent dryness. The fat in high-fat (marbled) meats can help contribute to juiciness and make the meat come out less dry when seared. High-collagen cuts of meat, like beef chuck, brisket, and flank steak, can become juicy when slow cooked for long periods, as the collagen transforms into gelatin, which is perceived as moistness.

Q DO I NEED TO LET MEAT AND POULTRY REST AFTER COOKING?

THE ANSWER No

THE SCIENCE In recipes for steak and other cuts of meat, you may see an instruction to let the meat rest for 5 to 10 minutes before serving or cutting it into pieces. The thinking behind that step is that moisture is pushed out to the surface as the meat cooks and the muscle fibers contract—so, if you cut into the meat as soon as it comes off the heat, the juices will come spilling out. Resting the meat is supposed to allow time for the fibers to relax and the juices to be reabsorbed and redistributed throughout the meat, resulting in a juicier steak. That all sounds great, but the reality is otherwise. First, as the meat rests, it continues to cook from what is known as carryover heat (see Is Carryover Heat Really a Thing?, page 64); depending on the size of the cut and how long you wait, the temperature can climb another 5°F to 10°F, possibly making for a dry result. Also, as the meat cools, your crust or your crispy chicken skin can soften, and delicious fats can start to solidify, affecting taste and texture. Finally, in several informal experiments conducted by meat scientists and chefs, it was found that the difference in moisture between rested and unrested steaks ranged between 6 to 15 percent, a relatively negligible amount.

KITCHEN TAKEAWAY Serve meat immediately after its been cooked to keep the meat hot when serving and prevent overcooking from carryover heat.

AS THE MEAT RESTS, IT CONTINUES TO COOK

Is Carryover Heat Really a Thing?

THE ANSWER Yes

THE SCIENCE When food is cooked, its exterior is always hotter than its interior because the surface has been more exposed to the heat source (except when microwaving, where heat is generated from the inside of the food). During the cooking process on a stovetop, in the oven, or on the grill, heat is transferred from the much hotter exterior to the interior of the food to establish thermal equilibrium. After the food is removed from the heat source, heat continues to redistribute from the exterior to the interior of the food. This phenomenon is known as carryover heat, and depending on certain factors, it can increase the interior temperature of a food by 3°F to 15°F, possibly affecting the moistness, texture, and flavor of the food as it rests away from its heat source.

The factors that affect carryover heat are external temperature (how hot is the pan, oven, or grill?), the size of the food item (mass), its water content, and the surface area of the food. Hotter surfaces will result in more heat moving from the outside in. Food that has been roasting at 450°F/232°C is going to have much more carryover heat than food that has been cooked for a long time at 250°F/121°C. Large foods like roasted turkey can retain quite a lot more heat than, say, roasted baby potatoes. The water content matters because water has a high heat capacity compared with fat or protein; even small amounts of water can hold quite a lot of heat, which means that high-moisture foods will retain more carryover heat than drier foods will. Surface area also plays a role; a whole potato will retain more heat than the same amount of sliced potatoes, since heat will quickly disperse from all the cut surfaces.

KITCHEN TAKEAWAY If you don't intend to serve a piece of meat as soon as it comes out of the oven or off the grill, stop cooking it when the internal temperature is about 5°F (for thinner cuts cooked at medium heat) to 10°F (for large pieces cooked at high heat) lower than the target temperature. Otherwise, cook the meat to your desired degree of doneness, then immediately slice, serve, and enjoy!

WHY ARE YOU SUPPOSED TO SLICE CERTAIN MEATS AGAINST THE GRAIN?

THE SCIENCE Meat contains muscle fibers that run parallel to one another. The more exercised these muscles are by the animal, the tougher and more rubbery these fibers become. If meat is sliced along the grain, or in the direction that the muscle fibers run, these long strands of muscle fibers remain in every bite. It doesn't matter how you cut meat from a part of the animal that doesn't get much exercise, like the tenderloin, since the muscle fibers in those cuts are not as dense, but if you cut a tougher steak or roast like flank steak or brisket along the grain, it will take quite a bit of unpleasant chewing to finally break those fibers down. When meat is sliced against the grain, the muscle fibers are fragmented and shortened, making it much easier to chew the meat.

America's Test Kitchen tested the effect of cutting against the grain versus with the grain on steaks using a texture analyzer to measure the amount of force required to chew the sliced meat. They found that when flank steak and strip loin were cut along the grain, the flank steak was four times as tough to chew as the strip loin was. However, when the two steaks were cut against the grain, the flank steak only required 16 percent more chewing force than the strip loin did. The results of this experiment are quite surprising and really show the power of slicing against the grain.

KITCHEN TAKEAWAY Slicing meats against the grain helps to increase their tenderness by breaking apart the tough muscle fibers. If sliced properly, even tough cuts of meat can become tender. Just take a moment to visually locate the grain, then slice the meat at an angle (up to 90 degrees) into slices; for additional tenderness, cut your slices as thin as you can.

IS THERE REALLY ANY DIFFERENCE BETWEEN REGULAR BROTH, BONE BROTH, AND STOCK?

ANSWER Yes

THE SCIENCE A broth or stock is great to have on hand in the kitchen when preparing soups, sauces, and stews. But what are the differences between them, and are they interchangeable? Broths and stocks are essentially flavorful solutions of the water-soluble components of meat or bones, vegetables, and aromatics. Regular broth is prepared by simmering meat (either with or without bones) in water for a short time, usually 30 to 60 minutes. That is enough time for some of the more soluble compounds in the meat to dissolve into the water, including umami-rich nucleosides, amino acids, meat flavor compounds, and some collagen (if you've included bones). Fats from the meat will also melt into the broth. When chilled, broth will remain fluid.

Stock, on the other hand, is prepared by simmering meaty, collagen-rich bones, like the leg or shank, oxtails, marrow bones, neck, thighs, and/or wings, in water for a longer period. Time is key to making a good stock; you need to simmer it long enough to extract and hydrolyze the collagen attached to the bones. This yields gelatin, which gives good stocks their silky, rich mouthfeel. When chilled, a well-made stock will become semisolid (like Jell-O) because the gelatin it contains will solidify as it cools, a process that will take anywhere from 2 hours (for chicken stock) to 6 hours. As the stock chills, any fat that melted into the stock while it simmered will rise to the top and solidify, forming a cap that can be easily removed.

Bone broth is similar to a stock in that it is simmered longer than a broth, but it takes that simmering time much further, to at least 10 to 12 hours. That is enough time not only to hydrolyze the collagen, but also to dissolve many of the mineral elements of the bone and extract as many nutrients from the bones as possible. When bone broth is refrigerated, it should also turn into a semisolid because of its gelatin content.

KITCHEN TAKEAWAY Broth, bone broth, and stock are all interchangeable in most recipes. Stock and bone broths are better used for sauces to lend them a thicker texture, and they can give soups and stews a richer flavor because of their gelatin content.

Does Citrus Juice Really Cook Seafood in Ceviche?

THE ANSWER No

THE SCIENCE You've probably noticed that when ceviche is prepared by marinating raw fish or other seafood in citrus juice, the seafood will appear to "cook" as the color of the flesh slowly changes from translucent pink to opaque white. This occurs because the proteins in the fish are changing their structure in response to the increased acidity from the citrus juice. The process is called denaturation and also happens when proteins are heated, which is why the same color change occurs when you sizzle a slab of fish on the grill. The trouble is that, unlike the fish sizzling on the grill, the raw fish marinating in citrus juice isn't being cooked in any scientific sense. While the citrus juice causes the color and texture of the fish to change, any bacteria or virus still lurking on the surface of the fish, and any parasites living inside the fish, will remain alive and well because no heat has been introduced to kill them off.

KITCHEN TAKEAWAY From a food safety standpoint, the fish in ceviche is still raw, so make sure you treat the fish the same way as you would when making sushi or other raw fish preparations. The fish should be absolutely fresh from a trusted vendor. Ceviche should be kept refrigerated while marinating, and you should eat it as soon as it's ready. Also, beware that ceviche left too long in the acidic marinade (more than 30 minutes) can become "overcooked" and break apart into chalky, dry pieces (see Can a Marinade Tenderize Meat?, page 54).

WHAT MAKES LOBSTERS TURN RED AND SHRIMP TURN PINK WHEN COOKED?

THE SCIENCE Astaxanthin is a red pigment molecule naturally produced by the microalgae *Haematococcus pluvialis* when the organism is stressed. The pigment is designed to protect the microalgae from the damaging effects of sunlight. Many sea creatures feed on this algae and accumulate the pigment in their tissues. Some fish, like salmon and red trout, have pinkish-red flesh because of their dietary intake of astaxanthin.

Lobsters and shrimp, on the other hand, produce a protein in their exoskeletons called crustacyanin. This protein binds to astaxanthin in a way that produces a blue color and is responsible for the blue-gray color of living lobsters and shrimp. When lobsters and shrimp are cooked, the crustacyanin protein loosens its hold on the astaxanthin and releases it into the surrounding shell. As a result, the red color is regenerated in the form of cooked red-pink seafood.

KITCHEN TAKEAWAY When a lobster or shrimp turns red or pink from cooking, it's a good sign that it's done, since the temperature at which crustacyanin loosens its hold on the astaxanthin, resulting in the color change of the seafood, is close to the denaturation temperature of the proteins in bacteria, which must be reached in order for lobster or shrimp to be safe to eat.

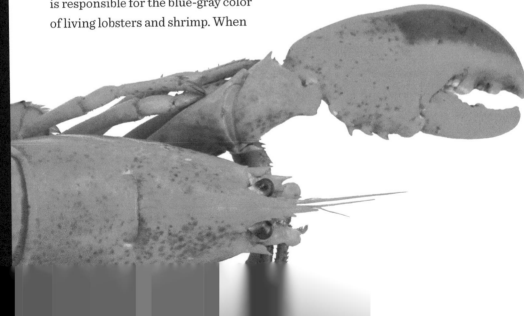

What Makes Fish Smell Fishy?

THE SCIENCE Fish contain a compound known as triethylamine oxide, which helps regulate the balance of saltiness between the ocean and their bodies. When fish are caught and killed, their tissue begins to break down. Fish tissue and symbiotic bacteria living inside the fish release an enzyme that converts triethylamine oxide into triethylamine, a highly volatile compound that is largely responsible for fishy odors. Humans are relatively sensitive to amines in general and have evolved to register amines as an indication of decay.

Fish also contain a lot of proteins, which are slowly broken down into amino acids by symbiotic bacteria and decaying fish tissue. The bacteria release a series of enzymes known as decarboxylases that break off carbon dioxide from the amino acid chemical structure, leaving behind amines like putrescine and cadaverine (named after the fact that they are the primary culprits behind the smell of rotting carcasses) that also contribute to fishy odors and flavors. These symbiotic bacteria proliferate if the fish is old or improperly stored at warmer temperatures, which is why spoiled or rotting fish has such a strong odor. Not only are the odors of amines particularly offensive, but amines can also be toxic if ingested in large quantities. One amine associated with a type of fish food poisoning is histamine, which can trigger itchiness, headaches, and diarrhea.

KITCHEN TAKEAWAY Amines react with acids to form ammonium salts, which are relatively odorless and flavorless. If you want to get rid of fishy odors in the kitchen, add a few tablespoons of vinegar to a pot of boiling water, which will vaporize the acetic acid in the vinegar and react with the odorous amines in the air. If a fish has a very strong smell, it is not fit for consumption. Any seafood you buy should not smell fishy; it really shouldn't have any smell at all.

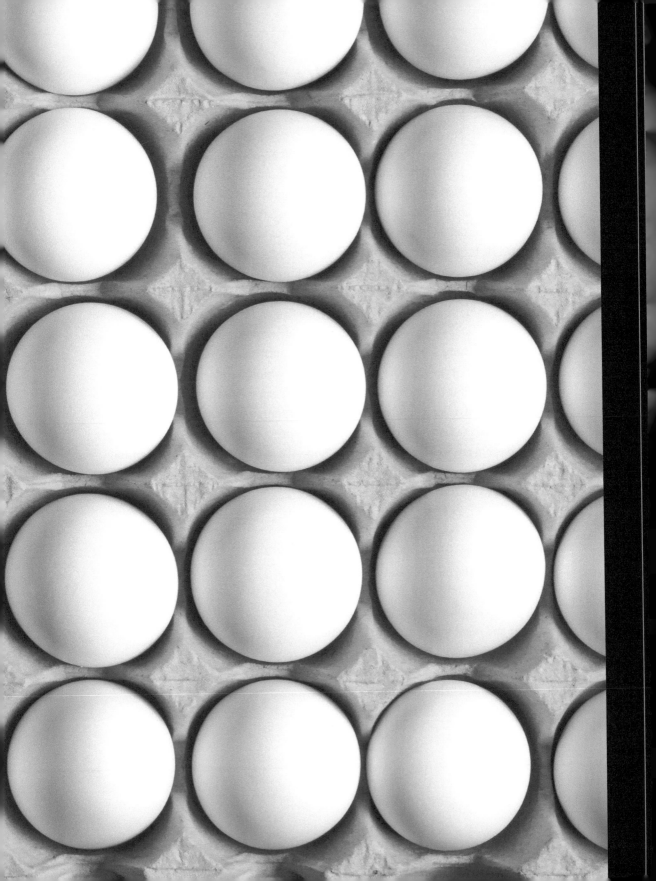

CHAPTER 4

Eggs and Dairy

EGGS AND DAIRY ARE THE FOUNDATION OF MANY FOODS, FROM RICH BREAKFAST DISHES TO TENDER, LIGHT-AS-A-FEATHER CAKES. The magic behind them is found in their miraculous, versatile proteins, which can create foams, emulsify oil and water, thicken batters, solidify into gels, and so much more! The fats found in eggs and dairy also lend creaminess, viscosity, and flavor to a variety of dishes. The ability to control the texture of an egg or dairy-rich dish is truly driven by the chemistry behind these fats and proteins. Let's dig deeper into the science of eggs and dairy.

IS THERE ONE BEST SUBSTITUTE FOR EGGS?

THE ANSWER No

THE SCIENCE Eggs are nature's most versatile source of protein and can serve any of multiple functions in a particular preparation—they can be whipped up into a foam, or they can be used to bind ingredients, thicken custards, emulsify sauces, and more. Choosing a plant-based alternative can be challenging because one size doesn't fit all—the substitute needs to be able to function in the same way that the egg does in the recipe.

Eggs are watery mixtures of proteins and fat, with the egg whites providing most of the moisture and thickening power, and the egg yolks providing the fat and emulsifying ability. The whites contain albumin proteins that form a network of interconnected proteins when whipped, trapping air and water to form first a foam and then a thick gel that gives lightness and airiness to foods. Egg white proteins also help bind fats, carbohydrates, and other proteins together, then coagulate and set when heated, which gives strength to baked foods.

The egg yolk is rich in fat and contains nearly half of the egg's total protein. Egg yolks contain lecithin, which is the primary emulsifier that gives the yolk its ability to emulsify fat and water-containing ingredients together.

Egg substitutes need to contain compounds that can simulate the effects eggs have in a specific recipe. The most common egg substitutes used in the kitchen are flax seeds, chia seeds, and aquafaba (chickpea soaking or cooking water), and these alternatives are typically used as binders or foaming agents, since they all contain compounds that form water-based networks when mixed with a batter or other ingredients. Flax seeds and chia seeds contain large, complex carbohydrates that trap water in a molecular network and cause them to swell. This can thicken a mixture in such a way that it mimics the gelling effect of eggs. Aquafaba contains a mixture of proteins and carbohydrates, so it can be added to a batter to function as a gelling agent or whipped up into a meringue-like foam.

well. If you need the foaming effects of egg whites in a meringue, mousse, or airy batter, aquafaba (3 tablespoons for 1 large egg) beaten until peaks form is your best bet. If you need the emulsifying effects of egg yolks, silken tofu, which contains the emul-

EGGS ARE NATURE'S
MOST VERSATILE
SOURCE OF PROTEIN

KITCHEN TAKEAWAY Eggs are complex mixtures of water, fats, proteins, and emulsifiers that have unique properties. Before replacing eggs with a plant-based substitute, you should know what function the eggs perform in the recipe—to create a foam, bind ingredients together, or emulsify? Each different plant-based egg substitute can usually only do one of these tasks

sifier soy lecithin, is a good choice (¼ cup for 1 egg yolk). If you require the creamy, gelatinous effect of whole eggs in a batter, replace them with a mixture of ground chia or flax seeds and water (1 tablespoon of ground seeds plus 3 tablespoons of water for each large egg); let the mixture sit for 15 to 20 minutes before using so the ground seeds can fully hydrate.

Why Do Egg Whites Expand When Beaten?

THE SCIENCE The transformation of egg whites into a billowing foam is a fascinating process. Ovalbumin is the main protein found in egg whites and is normally spherical in shape. When ovalbumin is disturbed by whisking, the mechanical force causes it to straighten out into strings. At the same time, oxygen from the air is being forced into the egg whites. This causes certain chemical bonds in the ovalbumin, called disulfide bridges, to break apart and find new partners. As more and more oxygen reacts with the proteins and rearranges these disulfide bridges, the stringlike proteins start to link together into a mesh-like network. Air becomes trapped and forms bubbles in this large protein network, and if a stabilizer has been added to the mix (such as salt, cream of tartar, or lemon juice), it strengthens the protein network by slightly denaturing the proteins, allowing them to unravel and reconnect more easily. Sugar can also improve the egg protein network by holding onto moisture and preventing the proteins from drying out at the surface of the foam air bubbles—if the proteins dry out, their structure will crumble and the foam will collapse.

However, there's a limit to this expansion. Too much beating can cause multiple bridges to form between neighboring proteins, aggregating into little clumps. This breaks down the protein network over time and results in a useless gritty mess that can't be recovered.

Fat can also disrupt the formation of the protein network, which is why it's important to make sure no residual egg yolk (which is rich in fat) gets mixed in with the egg whites. The fat molecules in the yolk contain water-loving and water-repelling chemical groups that will compete with the proteins surrounding the dissolved air bubbles, poking holes in the network and allowing the air bubbles to rapidly escape.

KITCHEN TAKEAWAY For the best results, your egg whites shouldn't contain even a speck of egg yolk. If you are separating a great number of eggs (for angel food cake, for instance), separate each egg into two small bowls. When you are sure your white is yolk free, transfer it to a larger bowl—the last thing you want to do is drop a bit of egg yolk into a big bowl of egg whites. Also, make sure your beaters and bowls are squeaky clean, free from any fat or soap residue that could affect the egg white foam.

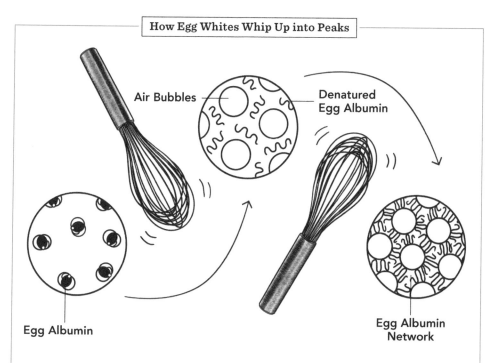

How Egg Whites Whip Up into Peaks

Air Bubbles

Denatured Egg Albumin

Egg Albumin

Egg Albumin Network

Egg whites contain the protein ovalbumin, which is normally globular in shape. When egg whites are whisked, the proteins unravel (denature) and air bubbles become trapped between them. With further whisking, the proteins start to bind to one another and create a mesh-like network that holds the air bubbles in place, causing the whites to, at the same time, lighten and thicken into a foam that can maintain peaks.

DOES TEMPERATURE MATTER WHEN BEATING EGG WHITES?

THE ANSWER Yes

THE SCIENCE Beating egg whites causes air to mix with the egg proteins (see Why Do Egg Whites Expand When Beaten?, page 74). Oxygen, coupled with mechanical agitation, causes these spherical proteins to straighten out and link up into a protein network. The speed at which these changes occur is dependent on temperature; the colder the temperature, the longer it takes. As such, room temperature egg whites beaten for the same length of time as egg whites cold from the refrigerator will achieve a greater volume.

KITCHEN TAKEAWAY Egg whites should be left to stand at room temperature for 30 minutes or placed in a warm bowl before beating to maximize foam volume. However, make sure you separate the egg yolks from the egg whites straight out of the fridge, because it's easier to do this when the eggs are cold.

Does Beating Egg Whites in a Copper Bowl Improve Volume?

THE ANSWER Yes

THE SCIENCE Ovalbumin, the main protein in egg whites, contains an amino acid called cysteine. Cysteine is rich in sulfur and can bond with other cysteines to form disulfide bridges. A small number of these disulfide bridges are desirable when beating egg whites because they help to form the protein network that traps air bubbles and forms foam. But too much beating can cause too many bridges to form, resulting in clumping and overbeaten egg whites that need to be thrown out.

When egg whites are beaten in a copper bowl, microscopic fragments of copper break off the inside of the bowl and dissolve into the egg whites to form copper ions. The sulfur in cysteine binds very strongly to copper, even more strongly than to other cysteines. These copper ions help keep too many cysteine disulfide bridges from forming in the beaten egg whites, and they also help stabilize the protein network. The egg whites stay fluffy and firm without the risk of becoming overbeaten—it is almost impossible to over-beat eggs in a copper bowl!

KITCHEN TAKEAWAY Egg white foams can be stabilized by beating them in a copper bowl. To maximize the power of a copper bowl, it should be freshly scrubbed with 1 tablespoon of salt and 1 tablespoon of vinegar or lemon juice to remove any oxidized copper on the surface. Copper oxides seal the surface of copper bowls that are meant for mixing food, preventing contact between the egg whites and fresh elemental copper.

Q HOW DO EGG WHITES CLARIFY STOCK?

THE SCIENCE Why does a stock become cloudy in the first place? It's usually the result of protein particles from the simmering meat that have either dissolved and then coagulated together or emulsified with melted fat floating in the stock to form rafts. Or, the cloudiness can come from particles that break off from the meat, vegetables, and spices added to the stock. Most of those particles represent flavor in one form or other, so there is no reason to remove them, other than for the sake of aesthetics.

While skimming and filtering a cloudy stock through cheesecloth can get rid of most the larger particles, clarifying it with egg whites helps remove the last remaining finer particles. So how does this magic work?

When egg whites are added to a warm stock, the egg albumin proteins dissolve in the stock and bind to denatured proteins, fat, and any other impurities. When the egg whites coagulate from the heat, these impurities become trapped in the albumin protein matrix. The coagulated egg whites are then filtered out to leave behind a crystal-clear stock.

KITCHEN TAKEAWAY You can clarify a stock after it's been prepared by whisking together two egg whites for every quart of stock until frothy. Whisk the egg whites into the warm stock, then bring the stock back to a gentle simmer for 15 minutes without stirring. A crust of solidified egg whites should form on the surface of the stock. Carefully remove the crust, and you should have a clear stock.

WHAT CAUSES THE SHELL TO STICK TO MY HARD-BOILED EGG WHEN I TRY TO PEEL IT?

THE SCIENCE The membrane of a freshly laid egg binds tightly to the shell and can make peeling nearly impossible. Older commercially sold eggs (as opposed to those sold direct from the farm) are usually easier to peel. This is because most commercially sold eggs are pre-washed by producers, during which a protective coating that surrounds the porous shell is removed, allowing carbon dioxide to seep out of the egg. Carbon dioxide is slightly acidic, so its release results in the egg becoming more alkaline, which makes it easier to peel. It's best to let eggs slated for hard-boiling sit in the fridge for a few days, where their pH can rise from 7.6 to 9.2 over time. Plus, during this sitting time, a little water will evaporate from the eggs through the shell, and the insides of the eggs will contract. This will cause the membranes to shrink inside the eggs, and air pockets will form between the shells and the eggs, also making them easier to peel.

KITCHEN TAKEAWAY If using farm-fresh eggs for hard boiling, wash them when you get home to remove their protective coating, and let them sit in the fridge a few days before boiling them for easier peeling. Alternatively, if you want to use farm-fresh eggs right away, add a half teaspoon of baking soda for every 4 cups of boiling water to increase the pH of the eggs while they boil.

Why Do Egg Yolks Sometimes Turn Green When Hard-Boiled?

THE SCIENCE When an egg is boiled for too long, the heat causes the sulfur atom in the amino acid cysteine found in egg white proteins to split off and form hydrogen sulfide. The hydrogen sulfide diffuses into the egg yolk, reacts with the iron bound there, and forms iron sulfide. The result is that dull green-black discoloration that can be found surrounding a boiled egg yolk. The yolk contains 95 times more iron than the egg white, which is why the egg white does not turn green. The green color is more likely to happen when older eggs are boiled; as an egg becomes older, it becomes more alkaline. When an older egg is cooked, these alkaline conditions accelerate the breakdown of cysteine and the production of hydrogen sulfide. Also, if you use a cast-iron pan when frying or scrambling eggs, iron can sometimes leach into the cooked eggs and form the same green discoloration.

THE GREEN COLOR IS MORE LIKELY TO HAPPEN WHEN OLDER EGGS ARE BOILED

KITCHEN TAKEAWAY To prevent hard-boiled eggs from forming a green ring, eggs should be brought to boiling from cold water and boiled for no more than 15 minutes. Then, the just-cooked eggs should be cooled down quickly with cold running water or an ice water bath to prevent overcooking. For scrambled eggs, whisk a few drops of lemon juice together with the eggs. The citric acid will bind to any residual iron and prevent the green color from forming.

Q WHAT MAKES FOR THE BEST HARD-BOILED EGG?

THE SCIENCE Steaming is a foolproof way to make flawlessly "boiled" eggs that are easy to peel. The reason a hard-boiled egg can be difficult to peel is that the membrane between the shell and the egg white is still cemented to the shell. When an egg is exposed to steam, the proteins in the membrane are heated quickly, causing them to rapidly break down (denature) and shrink away from the shell. Steaming also gives consistent, even results because the cold eggs don't touch the boiling water and cause the water temperature to suddenly drop.

KITCHEN TAKEAWAY Bring one inch (2.5 cm) of water to a boil in a saucepan. Set a steamer basket with a single layer of eggs in the saucepan, so the bottom of the steamer hovers above the water, then cover the pan and steam the eggs for 6 minutes for soft-boiled eggs or up to 13 minutes for hard-boiled eggs. Immediately place the eggs in a bowl of ice water, and let them chill for 15 minutes to stop the cooking. Peel the eggs under a stream of water to help release the membranes and shells.

Why Won't My Mayonnaise Thicken?

THE SCIENCE Mayonnaise is an emulsion, and emulsions can be easily disrupted. In the case of mayonnaise, the presence of egg yolk (which contains the emulsifier lecithin; see Why Are Emulsions Prone to Breaking?, page 20) is key to successfully combining oil and water; if your mayonnaise is not thickening, it could be that there isn't sufficient lecithin present and you should add more egg yolk to help bind the oil droplets to the water during mixing.

Applying enough force when whisking is also important, as it takes a lot of energy to force the dispersal of the oil into the water or the water into the oil. It's also important to add the oil to the yolk mixture very slowly while whisking rapidly to ensure the oil is broken into tiny droplets fast enough to disperse evenly throughout the mixture; if any large droplets remain (which can happen if you add the oil too fast), the smaller droplets can coalesce around the large ones and break the emulsion. Also, adding too much salt or too much acid (like lemon juice) can impede the creation of (or break) an emulsion by changing the solubility of the emulsifier.

KITCHEN TAKEAWAY Set yourself up for success when making homemade mayonnaise. First, follow the recipe exactly, adding the right amount of salt and/or lemon juice or vinegar to the egg yolk. Second, make sure you add the oil in a very slow stream (almost drop by drop), whisking very rapidly until the mixture thickens (emulsifies), at which point you can stream the oil in a bit faster (but don't dump it all in at once). If you have followed these directions and your mayo still is not coming together, you can add one to two teaspoons of boiling hot water to help re-dissolve the lecithin and set the egg yolks. You can also whisk in an extra egg yolk to add more lecithin and emulsifying power.

Q WHY DID MY HOLLANDAISE SAUCE BREAK?

THE SCIENCE When making hollandaise, you get a double shot of emulsifying power from both egg yolks (which contain lecithin) and clarified butter (which contains casein). The issues at play for making hollandaise are largely the same as for mayonnaise (see Why Won't My Mayonnaise Thicken?, page 80), so follow those Kitchen Takeaway guidelines here, too. The big difference that sets hollandaise sauce apart from mayo is that it is a cooked emulsion. Heating an emulsion can cause the dispersed oil droplets to expand and decrease in density, which results in them separating from the water, which has a higher density than the oil. Because of the delicate balance of this emulsion, hollandaise sauce is prepared using a double boiler over simmering water, rather than a hot pan over direct heat, and the sauce is never reheated.

Another factor that can cause hollandaise sauce to break is cold temperatures. Freezing can cause the water phase to form ice and disrupt the fragile emulsifier coating surrounding the dispersed oil droplets. As a result, when the emulsion is defrosted, it will separate.

KITCHEN TAKEAWAY If you are not going to use it immediately, pour the hollandaise sauce into a thermos to keep it hot until you are ready to serve. Unfortunately, if you overheat a hollandaise to the point of separation, the only thing you can do is throw it out and start again. Emulsions that have separated because they've been frozen are likewise unsalvageable.

WHY DOES A SKIN FORM ON CUSTARD OR HEATED MILK?

THE SCIENCE There are two types of milk proteins: casein and whey protein. Heating milk to 158°F/70°C or higher causes these proteins to rearrange their molecular structure in such a way that they bind to one another. This semi-permanent structural change is known as denaturation. Heating the milk also causes evaporation at its surface, concentrating the denatured proteins there. Once the concentration of whey and casein proteins exceeds a certain threshold, a skin begins to form. Additionally, the fat in milk binds to sections of the denatured proteins, which prevents them from re-dissolving back into the milk and forces them to the surface. A skin can also form when milk or a milk-containing mixture is heated and cooled down, because its surface will continue to evaporate during the cooling process.

That being said, there is nothing wrong with the milk skin, which is edible.

KITCHEN TAKEAWAY To prevent a skin from forming, the milk or milk-containing dish should be heated below 158°F/70°C and stirred constantly. Covering the pot also helps by minimizing surface evaporation. To prevent a skin from forming on a dairy-based pudding or custard, as soon as you transfer it to a bowl, place a sheet of plastic wrap directly on its surface to stop evaporation.

THERE IS NOTHING WRONG WITH THE MILK SKIN, WHICH IS EDIBLE

Can I Substitute Milk for Half-and-Half or Heavy Cream, and Vice Versa?

THE ANSWER Yes and no

THE SCIENCE When raw milk first comes into a milk processing plant, it's spun through a conical separator to split the contents into cream (which is 36 to 40 percent milk fat, 2.5 percent milk protein), and skim milk. The cream is packaged as heavy cream, used to produce butter, or added back to the skim milk at precise fat percentages to produce 1 percent milk, 2 percent milk, whole milk (which is 3.5 percent fat), and half-and-half. This process ensures that all milk and dairy products, regardless of which farm or what breed of cow the milk comes from, are standardized to the proper percentage of fat.

The fat content found in these products isn't the only thing that differentiates them. Since skim milk is the basis of 1 percent milk, 2 percent milk, whole milk, and half-and-half, the milk protein content in skim milk is important. Milk protein acts as an emulsifier that binds to other fat components in a recipe, which is essential for texture, thickening, and foaming. If you dilute heavy cream or half-and-half with water to make "milk," you'll be missing the proper ratio of these proteins that help with texture and mouthfeel. On the other hand, heavy cream and half-and-half can be replaced with a mixture of whole milk and butter, which makes up for the missing fat in whole milk.

KITCHEN TAKEAWAY Substitute ¾ cup of milk and ⅓ cup of butter for 1 cup of heavy cream (even if you're using it to make whipped cream), and substitute ⅞ cup of milk plus ½ tablespoon of butter for 1 cup of half-and-half.

Q DOES BUTTERMILK CONTAIN BUTTER?

THE ANSWER No

THE SCIENCE Butter was first produced by allowing milk to sour overnight, which caused the cream to separate out. The soured milk was then churned into butter and the liquid that is left over from churning was given the name buttermilk. Despite its name, buttermilk has less fat than milk because all the fat is removed in the churning process. Today, buttermilk is made industrially by culturing low-fat or skim milk with commercial bacterial cultures for 12 hours. The bacteria ferment lactose into lactic acid, which gives buttermilk its sour taste. The increased acidity causes the milk proteins to precipitate out of the milk, which in turn coagulate and thicken the milk. When buttermilk is added to batter in the presence of baking soda, the lactic acid reacts vigorously with it to produce a different rise and texture than other leaveners, which is why we love our buttermilk pancakes, waffles, cornbread, and biscuits.

KITCHEN TAKEAWAY Buttermilk is used to help leaven batters and baked goods through its quick and unique reaction with baking soda. Buttermilk can also be used in place of milk to add a bit of tanginess to soups, ice cream, salad dressing, mashed potatoes, and oatmeal.

Why Do Different Cheeses Melt Differently?

THE SCIENCE Cheese is composed of water, milk proteins, fats, lactose, and calcium ions (sounds appetizing, doesn't it?) and is held together by bridges of calcium ions that link to casein, the main protein in cheese. Inside the voids of this casein network reside the water, fat, and lactose. When cheese is heated, the solidified milk fats first liquefy, then separate from this cheese matrix. With continued heating, the casein proteins break free from the calcium ions and begin to move freely. The temperature at which this separation takes place and the amount of freedom with which the casein moves (the viscosity) varies with different cheeses.

Several factors affect the melting temperature and viscosity of cheese. The cheese's moisture content is one such factor. In high-moisture cheeses like mozzarella and Brie, the water acts like a lubricant between the proteins, resulting in a lower melting temperature. Fat content plays a similar role in the melting properties of a cheese. High amounts of fat (in cheeses like Cheshire and Leicester) result in what you might call melty cheeses that won't hold their shape. Another factor is age. Young cheeses like queso fresco and provolone contain intact casein molecules that hold the casein network together and form a bouncy, stringy texture that remains even when they are melted. Some young cheeses, like Halloumi and paneer, contain such a high level of protein and low level of fat that they resist melting because these protein networks keep the cheese intact even at high temperatures. As a cheese ages, protein-degrading enzymes released by bacteria and mold fragment the casein proteins and help to create a smoother melting cheese, like sharp aged cheddar, for example.

KITCHEN TAKEAWAY All cheeses are not the same when it comes to melting, which is a great thing—you can always find a cheese that suits your need, whether you want to grill it or turn it into mac and cheese. To transform any cheese into a smooth, melty cheese sauce, one trick is to combine 2 teaspoons of citric acid (you can find it where canning supplies are sold), 2½ teaspoons of baking soda, ½ cup of water, and ½ pound of your favorite cheese, shredded; heat the mixture gently until the cheese is melted, then stir gently until the sauce is smooth. Citric acid and baking soda react to form sodium citrate, which binds to calcium and breaks apart the casein network.

Q WHY DO SOME CHEESES FORM RINDS AND OTHERS DON'T?

THE SCIENCE Bloomy, white rinds like the ones you find on Brie and Camembert are made up of edible molds from the *Penicillium* genus (yup, the same genus that brings you penicillin). To make these types of cheeses, cheesemakers spray a solution containing live cultures of the mold onto a cheese, then set the cheese in a humid environment that encourages the mold's growth. Other microorganisms will make their home in these rinds as the mold breaks down the surface of the cheese, creating the familiar white rind, which usually begins forming 5 to 12 days after its inoculation.

Washed rind cheeses have an orange or reddish hue, thanks to the growth of microbial communities that produce red, orange, and yellow pigments. These types of cheeses are washed daily with a brine or alcohol solution to encourage the growth of *Brevibacterium linens*, bacteria responsible for the strong flavor and odor of certain aged cheeses, such as Limburger and Appenzeller. Cheeses with what are called smear-ripened rinds form in a way that is similar to that of washed rind cheeses, except that the wash itself contains bacteria or fungi to add flavor. Muenster and Port Salut are smear-ripened rind cheeses.

Natural rinds form simply as a result of the exterior of a cheese drying over time. Cheeses without rinds, like mozzarella and cheddar cheese, are either too young to form rinds or commercially aged inside plastic wrapping to prevent the formation of rinds.

KITCHEN TAKEAWAY Most rinds are edible. Only rinds made from wax, like those found on Edam and Gouda, or rinds too tough to chew should be avoided. However, if a cheese rind is too tough to chew, consider saving it for cooking. Tough rinds from cheeses like Parmesan can be added to a pasta sauce, stew, or soup for flavor, and other tough rinds can be heated on the grill until they are soft enough to eat.

Why Do Some Cheeses Have Strong Aromas and Others Don't?

THE SCIENCE The production of most cheeses starts in a similar way. Cheese is produced by curdling milk with an acid or enzymes, heating and salting the solid curds, then used as is, forming the cheese into shapes, or pressing the curds into wheels or blocks. Freshly pressed cheese curds are quite mild in flavor and aroma. So why do some cheeses smell strongly while others remain mild? The difference lies in the types of bacteria and molds used to age each cheese and the length of the aging. Many smelly cheeses tend to be made using the washed-rind or smear-ripened method (see Why Do Some Cheeses Form Rinds and Others Don't?, page 86), both of which expose the surface of the cheese to moisture and high humidity, encouraging the growth of bacteria and molds that generate odorous compounds. Drier aging conditions tend to support the growth of milder-smelling microorganisms.

Let's take Limburger cheese as an example. Arguably one of the world's worst smelling cheeses, it is inoculated with the bacteria *Brevibacterium linens*, the same bacteria that's responsible for foot and body odor. Limburger is then washed with a mild brine once a day for several weeks and aged for several months. The bacteria metabolize sulfur-containing amino acids in the cheese proteins and produce noxious volatile sulfur compounds like methyl mercaptan and hydrogen sulfide—the same compounds that give cabbage and rotten eggs their terrible smell. *B. linens* also produces the putrid-smelling butyric and valeric acids, which are responsible for the odor of old gym clothes. It takes about three months for Limburger cheese to transform from fresh milk curd to the odor bomb that we know and some people love, and as you can see, there is quite a bit of biochemistry going on during that time.

KITCHEN TAKEAWAY Stinky cheeses can be a big, smelly problem when they are improperly stored in the fridge. The odor compounds are small molecules that can easily escape plastic bags or plastic wrap and be absorbed by other foods. They can also lose their flavors over time for the same reason. To preserve what you love best about your stinky cheese, take it out of its original wrapper, rewrap it in parchment or waxed paper, and then wrap it a second time in aluminum foil; finally, place it in a hard-plastic or glass storage container with a snap lid and store it in the refrigerator.

WHY DOESN'T THE MOLD IN BLUE CHEESE MAKE YOU SICK?

THE SCIENCE Blue cheese is inoculated with a mold known as *Penicillium roqueforti*, which is responsible for its characteristic flavor, odor, and blue veins. Historically, blue cheeses were naturally inoculated by *P. roqueforti* spores that were present in the air of cheese factories, but modern methods now use freeze-dried *P. roqueforti* spores that have been commercially manufactured in aseptic laboratory conditions. After the cheese has been inoculated with the mold, it is aged for 60 to 90 days at cool temperatures and humidity levels that control for spoilage. After this ripening period, the cheese is sterilized at 266°F/130°C for four seconds to kill any remaining *P. roqueforti* mold and prevent further fermentation.

KITCHEN TAKEAWAY While the thought of eating a mold may seem dangerous (or icky), there is nothing inherently dangerous about consuming blue cheese or other purposely moldy cheeses (which include Brie, with its white, bloomy rind). However, soft cheeses that aren't meant to be moldy and have developed mold in your refrigerator should definitely be thrown away, since the mold mycelium can penetrate into the cheese and will definitely make you quite sick for a couple of hours. Moldy spots from hard cheeses can simply be cut off because mold has a hard time growing deep into hard cheeses.

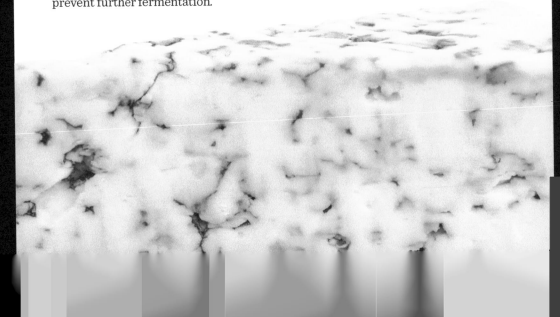

Q CAN I FREEZE CHEESE?

THE ANSWER It depends on the cheese

THE SCIENCE Cheese is a tasty combination of protein, fat, water, and salt. While cheeses vary in terms of their ratios of each of these components, it's their water content that most affects whether they can be frozen and then thawed with good results. When a food is frozen, the water molecules slow down and arrange themselves into a crystalline structure to form ice. Because ice crystals take up more volume than liquid water, frozen water expands. When the water in cheese freezes, it causes fissures and fractures to form in the cheese protein microstructure, resulting in textural defects. When the cheese is defrosted, the water is not reabsorbed into the protein-fat matrix; instead, it separates out from the cheese into microscopic pools, creating a more crumbly, granular cheese. Flavor is also affected by this separation, as water- and fat-soluble flavor components concentrate into the separated phases, rather than being more evenly distributed throughout the cheese.

Higher-moister cheeses like queso fresco, paneer, and Brie suffer the most from the effects of freezing. Delicate artisanal cheeses and cheeses with holes and pores are also negatively affected by freezing. However, most industrially produced block or shredded cheeses are designed to withstand freezing and will remain in relatively good condition after thawing. Hard, low-moisture cheeses like Parmigiano-Reggiano and Pecorino Romano are also less susceptible to the damaging effects of the freezer.

KITCHEN TAKEAWAY If you intend to freeze a cheese, wrap it properly to ensure best results. Tightly wrap the cheese in plastic wrap, then place it in a zip-top bag or hard-plastic or glass storage container with a snap lid; this should prevent freezer burn. Frozen cheeses will keep for up to three months and should be defrosted in the fridge overnight.

Is Processed Cheese Fake?

THE ANSWER No

THE SCIENCE James Kraft, the founder of Kraft Foods, was originally a wholesale cheese merchant back in the early 1900s, selling to individual grocers. While building up his business, he quickly realized that one of the challenges that prevented him from reaching more customers was the fact that cheese would spoil when transported across long distances. In 1916, he patented a process to manufacture a cheese that could resist spoilage. In the original patent, blocks of cheese, typically cheddar, Colby, Swiss, or provolone, were melted with sodium citrate or phosphate salts, which helped keep the fat and proteins from separating during the melting process. The liquid cheese was rapidly pasteurized to kill off the microorganisms responsible for spoilage, and then it was injected into sealable tin cans. The result was a cheese that could be shipped across the world and stored nearly indefinitely.

Processed cheese makers eventually discovered that processed cheese could be cut with cheaper ingredients like crude whey and dairy solids produced as waste products by normal cheese manufacturing. Preservatives, emulsifiers, food dyes, and artificial flavors were added to improve texture, flavor, and melting properties. But the real innovation came when developers found that they could add enzymes to a processed cheese and warm the mixture to optimized temperatures for several days, which allowed the enzymes to break apart the proteins and fats in the cheese and generate very strong flavor molecules. The result was a highly concentrated form of cheese flavor known as enzyme-modified cheese. This process has been engineered to reproduce the flavor of aged cheddar, Parmesan, Gorgonzola, and other strong cheeses in only a few short days instead of the months and years normally required. Enzyme-modified cheese is now added to food products like baked snacks, dips, soups, and other processed cheeses to intensify their flavor.

CHAPTER 5

Fruits and Vegetables

HUMANS HAVE INTERVENED IN THE EVOLUTION OF FRUITS AND VEGETABLES THROUGH SELECTIVE BREEDING SINCE THE INVENTION OF AGRICULTURE, WHICH HAS GIVEN US THE MANY VARIETIES WE ENJOY TODAY. We've long adapted the biochemistry of plants to satisfy our appetite, giving rise to sweeter apples, starchier potatoes, juicier peaches, and milder broccoli. But once we've brought back the day's shopping to the kitchen, there's still much to do to transform these ingredients to suit our palates. Relying on the science of food, we can maximize their flavors and textures for our gastronomic delight.

What Happens When Fruit Ripens?

THE SCIENCE When it comes to ripening, there are two varieties of fruits: climacteric and non-climacteric. Climacteric fruits continue to ripen after they've been picked; these include apples, bananas, tomatoes, and avocados. Non-climacteric fruits must remain on the plant to finish ripening.

When a fruit flower is first pollinated, the fruit seed becomes fertilized and begins the process of fruit growth. Seeds absorb moisture, nutrients, and sugars while releasing hormones that induce cell division in the ovary wall and cause the cells to expand in size to form an unripe fruit. The fruit contains a high amount of tannins, alkaloids, and dense fibers, all of which give unripe fruit its bitter, astringent flavor and tough exterior flesh. The purpose of these compounds is to prevent bacteria, fungi, and animals from prematurely eating the fruit and its underdeveloped seed. Once the fruit has reached full size, a series of genes are switched on and enzymes are produced by the fruit to convert the dense, unripened fruit into a juicy, succulent treat that's now attractive to animals and humans (who will eat it and then disperse the seeds).

The fruit begins respiring oxygen to generate energy and heat during this process. Starches turn into sugars, acids are neutralized, green chlorophyll is broken down into new pigment molecules, large molecules are turned into aromatic compounds that give ripe fruit its distinct aroma, and pectin fibers that keep the fruit cells glued together are hydrolyzed to form softer flesh. The process continues until bacteria and fungi begin to decompose the fruit into a mushy mess.

KITCHEN TAKEAWAY Many climacteric fruits have a small window of ripening, meaning that once they start ripening, they progress to overripe and mushy pretty fast. To slow down the enzymes involved in the ripening process, fruits like avocados and bananas can be kept cool in the fridge or in a cool part of the house.

WHY DO SOME FRUITS BROWN WHEN CUT?

THE SCIENCE This type of browning, initiated by enzymes collectively known as oxidases (and not surprisingly called enzymatic browning) is characteristic of many fruits, including bananas, avocados, and apples. The main culprits are polyphenol oxidase and catechol oxidase, which jump into action when exposed to oxygen, converting compounds found naturally in the fruit into melanin, which absorbs light and gives the cut fruit a dark, brownish color. Melanin is the same compound that gives humans darker skin tones when they are exposed to ultraviolet light. Both polyphenol oxidase and catechol oxidase function optimally at a neutral pH of 7, so acids can help slow down these enzymes and the subsequent browning.

KITCHEN TAKEAWAY To counteract browning, cover the cut fruit tightly with plastic wrap to limit its exposure to oxygen, which kick-starts the process, and/or coat the cut surfaces with lemon or lime juice (which has a pH of 2 to 3) to slow down the enzymes that cause the browning to take place.

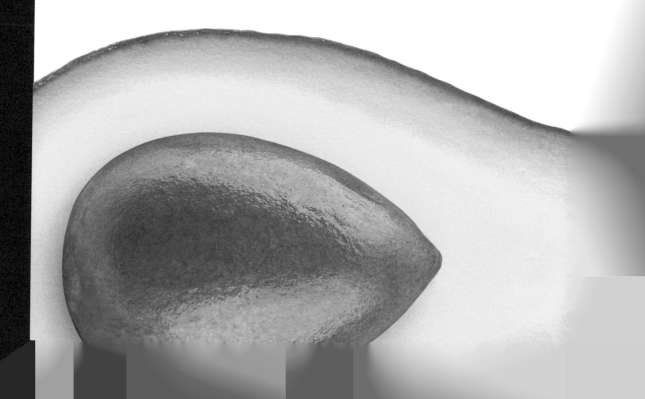

Does Putting a Banana with an Underripe Fruit Really Help the Fruit Ripen Faster?

THE ANSWER Yes

THE SCIENCE Certain types of fruit, known as climacteric fruits, can trigger themselves into ripening. When they reach the ripening stage, genes are switched on in the fruit that cause the production and release of ethylene gas. Climacteric fruits contain receptors that bind to ethylene gas and initiate a complex series of physiological processes that result in ripening. Bananas function in this way, as do apples and tomatoes. There is also a number of other fruits that are highly sensitive to the effects of ethylene gas, including lemons, limes, mangos, pears, peaches, and avocados. When these fruits are put in close proximity to one another, ripening can happen pretty quickly.

Controlling and producing ethylene gas is an important process for commercial agriculture. Ethylene gas is used to help ripen fruit after it has been transported long distances. Normally, farmers pick fruits when they're unripe so that they can be stored in warehouses and survive transit without decaying. During storage and transportation, these fruits are kept with ethylene scrubbers and absorbers to prevent premature ripening. Once the unripe fruit reaches its destination, the fruits are sprayed with synthetic ethylene gas to accelerate ripening before they're displayed on the grocery shelves. Ancient practices for ripening fruit also relied on ethylene gas, even though it was not yet understood that ethylene was the chemical responsible for these effects. Ancient farmers and harvesters would place slashed-open fruit together or burn incense in a room with fruit to increase ripening—later, it was discovered that slicing fruit stimulates ethylene production, and that burning incense generates minute levels of the gas as a by-product of combustion.

KITCHEN TAKEAWAY The classic trick of putting a banana in a brown paper bag with a piece of unripe fruit speeds up ripening by 1 to 2 days by trapping and concentrating the released ethylene gas.

Q DO CANNED AND FROZEN FRUITS AND VEGETABLES HAVE FEWER NUTRIENTS THAN FRESH PRODUCE?

THE ANSWER It depends

THE SCIENCE Fruits and vegetables that have been picked at their peak freshness will have the highest possible amount of nutrients. However, in the time it takes produce to be shipped and transported to the supermarket and then eaten, a substantial portion of the nutrients can be lost because fruits and vegetables are still living and continue to metabolize the nutrients found inside them. Canning and freezing can preserve these nutrients. Canning, which involves heating food products in a sealed glass or metal container to preserve them, deactivates enzymes naturally found in some fruits and vegetables that can degrade nutrients. Produce that is commercially frozen is first blanched, which also deactivates enzymes that can degrade nutrients.

KITCHEN TAKEAWAY There will always be some nutrient loss with fruits and vegetables, unless they're eaten fresh from the field. Frozen or canned fruits and vegetables can sometimes contain more nutrients than fresh produce that's been kept around for too long, so it's an excellent option if fresh isn't available or appealing.

THERE WILL ALWAYS BE SOME NUTRIENT LOSS
WITH FRUITS AND VEGETABLES
UNLESS THEY'RE EATEN
FRESH FROM THE FIELD

Do Fruits and Vegetables Lose Their Nutrients When Cooked?

THE ANSWER It depends

THE SCIENCE The cooking process applies heat to fruits and vegetables, and many nutrients are sensitive to heat. Vitamins, for example, can easily degrade in the presence of heat and oxygen. The way a food is cooked also influences nutrient levels. Water-soluble nutrients like vitamins B and C and minerals easily dissolve out of fruits and vegetables when they're boiled. Fat-soluble nutrients A, D, E, and K can dissolve in the cooking oil used for sautéing or frying. Omega-3 fatty acids are more prone to degradation when ingredients are fried rather than boiled. Steaming applies more gentle heat and can preserve most of the heat-sensitive vitamins in fruits and vegetables better than most other cooking methods.

At the same time, cooking helps breaks down the tissues in fruits and vegetables, which makes some nutrients more bioavailable. Potatoes and other starchy vegetables are rarely or never eaten raw because their energy-rich starches are trapped inside plant cell walls and are inaccessible to our digestive system unless cooked. Certain nutrients become easier to absorb when cooked, like the antioxidants beta-carotene and lycopene, which give carrots and tomatoes, respectively, their orange and red color.

KITCHEN TAKEAWAY The effects cooking has on the nutrient content and quality of fruits and vegetables are complex, and there is no simple answer to whether cooked produce is better or worse than raw. While some nutrients are broken down, others can be enhanced by the cooking process.

WHY DOES MY TONGUE TINGLE WHEN I EAT PINEAPPLE?

THE SCIENCE Pineapple contains an enzyme called bromelain that cuts the fruit's protein molecules into smaller fragments. When pineapple juice hits your mouth and tongue, bromelain acts on the surface proteins and begins digesting them slowly. As you continue to chew, bromelain will eventually create microscopic openings on your tongue, which the acids in the pineapple will react with to cause a stinging sensation. This chain of events will only occur when you eat fresh pineapple, because the cooking and processing of canned pineapple deactivate bromelain.

Papaya contains a similar protein-digesting enzyme, papain, but you won't experience the same stinging because the pH of papaya is neutral.

KITCHEN TAKEAWAY If you don't like that stinging feeling but you prefer to eat fresh pineapple rather than canned, try roasting, grilling, or pan-searing the cut fruit—these cooking methods will deactivate the bromelain and caramelize the sugars. You can also microwave fresh pineapple for 3 to 5 minutes to prevent the stinging sensation.

What Makes a Chile Pepper Hot?

THE SCIENCE Chiles contain a class of heat-inducing molecules known as capsaicinoids. Capsaicin and its molecular cousin dihydrocapsaicin are the spiciest of these molecules, with capsaicin making up a large proportion of the heat in chiles. The experience of spiciness and heat is subjectively measured on the Scoville scale (in Scoville heat units, or SHU), with pure capsaicin and dihydrocapsaicin each rated at 16 billion SHU. These molecules are fat-soluble compounds that interact with a receptor called TRPV1 in the sensory neurons found in our mouths, nostrils, and eyes. Capsaicin and capsaicinoids have precisely the right molecular structure to bind strongly to the TRPV1 receptor, which causes the neurons to send a signal to the brain that mimics the effect of pain in response to extreme heat or abrasion.

Incidentally, the ability of mammals to experience pain and heat when ingesting chiles seems to be a quirk in evolution, as birds possess the TRPV1 receptor but not the same neural pathway. Humans are the only mammals that enjoy eating chiles for the sake of the heat. For some, ingesting high levels of capsaicin can cause the experience of euphoria and pleasure in response to the pain through a mechanism called hedonic reversal. Paul Rozin, a professor of psychology at the University of Pennsylvania, studied the relationship humans have with foods that are spicy, bitter, or disgusting. He found that, over time, cultural upbringing and social pressure

FOR SOME, INGESTING HIGH LEVELS OF CAPSAICIN CAN CAUSE THE EXPERIENCE OF EUPHORIA AND PLEASURE

can create a kind of masochistic pleasure (hedonic reversal) when eating seemingly unpleasant foods, despite the body's negative reaction to the food.

KITCHEN TAKEAWAY There's a commonly held belief that the seeds are the hottest part of a chile, but this is not true. The highest concentration of capsaicin is found in the white pith surrounding the seeds, so that is the part to remove if you want to turn down the heat. Be careful not to touch your eyes or nose after handling chiles, because the capsaicin can bind to pain receptors found in mucus membranes and cause you a good deal of discomfort.

Q WHAT CAN YOU DO TO RELIEVE CHILE BURN?

THE SCIENCE Researchers at the University of California, Davis, conducted a study to determine the most effective way to stop chile burn. Since capsaicin is soluble in fat, they believed that ingesting or rinsing the mouth with liquids containing fat or fat-dissolving liquids like alcohol would be effective. It turned out that rinsing your mouth with ethanol was about as effective as rinsing it with water, meaning that it wasn't. Whole milk did reduce the heat, but upping the fat quotient by adding butter didn't increase its effectiveness. Surprisingly, the researcher found that the most effective way to relieve the burning sensation was to rinse your mouth out with a cold glass of sugar water. Though it's still unclear why this method works, the researchers hypothesized that the sugar may serve as a neural distraction to the brain that interferes with the perception of heat and pain from the chiles.

KITCHEN TAKEAWAY If you accidently chomped on a spicy chile and are experiencing overwhelming heat, go for a cold glass of sweet juice or soda. Milk will work, too, but anything with sugar will be more effective. If you get capsaicin in your eyes or nose, try flushing them with milk or sugar water to quiet the burn.

Why Does Cutting Onions Make You Cry?

THE SCIENCE Onions have evolved a special defense system that prevents animals from munching on them in the wild. When you chop, slice, or bruise an onion, the cell walls break apart and release two enzymes and an amino acid unique to this vegetable.

The first enzyme, known as alliinase, biochemically chops the amino acid in half and produces a compound called a sulfenic acid, which is the source of that piquant, sulfurous flavor shared by garlic, onions, shallots, leeks, chives, and other members of the allium family. But unlike garlic or leeks, a second enzyme, called lachrymatory factor synthase, quickly transforms that flavorful molecule into another chemical known to scientists as the lachrymatory factor. This is the chemical that causes your eyes to well up as you're dicing an onion (not surprisingly, *lachrymatory* means to tend to cause tears). And as you probably know firsthand, these enzymes act fast. Within minutes, 99% of the amino acid is immediately transformed into a burst of lachrymatory factor, leaving you hunting for a tissue.

KITCHEN TAKEAWAY If you aren't a fan of shedding tears in front of the chopping block, try this onion hack: Cut an onion in half, then microwave the halves for 2 to 3 minutes. Refrigerate them until they're cool to the touch, then mince or chop them. When you're done, mince a clove of garlic and mix it into the onion with 1 tablespoon of water; set aside for 5 minutes. The heat from the microwave will deactivate the onion enzymes, which unfortunately includes the flavor enzyme, but tossing the onion with the garlic and water will add the flavor enzyme back in.

The Biochemistry Behind Onions and Tears

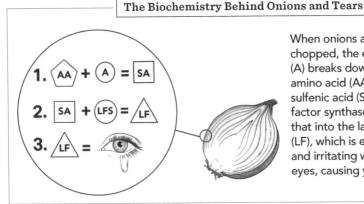

When onions are sliced or chopped, the enzyme alliinase (A) breaks down an onion-based amino acid (AA) and forms a sulfenic acid (SA). Lachrymatory factor synthase (LFS) transforms that into the lachrymatory factor (LF), which is extremely volatile and irritating when it gets in your eyes, causing you to tear up.

IS THERE ANY REAL DIFFERENCE BETWEEN YELLOW, RED, AND WHITE ONIONS?

THE ANSWER Yes

THE SCIENCE Onions are part of the genus *Allium*, which also includes similarly pungent vegetables like garlic, chives, leeks, and shallots. Alliums are unique because they have a built-in chemical defense system to ward off animals and pests. When the tissues of an allium are damaged, an enzyme called alliinase releases sulfur compounds (sulfenic acids), which give alliums their characteristic bite when eaten raw. Alliinase works by acting on a class of amino acids stored in allium tissues called cysteine sulfoxides, and each allium vegetable contains a different set and amount of cysteine sulfoxides.

Onions have a specialized chemical system that goes one step further than other alliums. When onions are damaged, alliinase is released from the cells and reacts with an onion-specific cysteine sulfoxide to form an onion-specific sulfenic acid. That sulfenic acid is transformed by a second enzyme called lachrymatory factor synthase to form what is known as the lachrymatory factor, which causes your eyes to water when an onion is chopped or sliced (for more on that, see page 102). White onions have high amounts of these enzymes and the onion-specific cysteine sulfoxide that gives them an intense bite and pungency. Yellow and red onions are bred to contain less of the onion-specific cysteine sulfoxide that serves as a precursor to the lachrymatory factor. Sweet yellow or Vidalia onions are bred to contain fewer alliinase enzymes, so little sulfenic acid or lachrymatory factor is produced when these onions are sliced, resulting in fewer tears and greater sweetness.

KITCHEN TAKEAWAY Different onion varieties are bred in different ways that affect their biochemistry and produce unique flavor profiles as a result. White onions have a more intense bite and oniony flavor; red and yellow onions are more mellow in flavor; and Vidalia, Maui, and other sweet varieties have largely had the pungency bred out of them. Choose your onion based on your own preferred level of oniony flavor and bite.

Fruits and Vegetables

HOW DO GARLIC AND ONION TRANSMIT FLAVOR TO A DISH?

THE SCIENCE Garlic and onion contain a special class of amino acids known as cysteine sulfoxides that can transmit flavor in several different ways. First, when an onion or clove of garlic is cut into, their special class of amino acids release sulfenic acid, which goes through a further transformation to become the flavors we recognize as onion and garlic. Second, when exposed to heat, cysteine sulfoxides will react with the sugars in garlic and onion to trigger the Maillard reaction, producing sulfur-containing aromas and flavor compounds that mimic those found in meat dishes. (Black garlic, whole garlic that is aged for four weeks at 120°F/60°C, is particularly rich in these Maillard products.)

Finally, if cysteine sulfoxides are combined with umami-producing compounds like monosodium glutamate, disodium inosinate, and disodium guanylate (which can be found in ingredients like anchovies, fish sauce, soy sauce, shiitake mushrooms, tomato sauce, yeast extract, Parmesan cheese, and meat bouillon), they will increase the intensity of the umami flavor and extend its duration. This taste property is known as kokumi (for more on kokumi, see What Gives Nutritional Yeast Its Distinctive Flavor?, page 44).

KITCHEN TAKEAWAY Different chemical reactions are at play, depending on whether garlic and onion are raw or cooked, but either way, they provide savory umami notes that add to the complexity of a dish.

Does the Way You Cut Up Garlic Affect How Strong It Is in a Dish?

THE ANSWER Yes

THE SCIENCE Garlic contains the enzyme alliinase, which acts on amino acids (cysteine sulfoxides) to release a sulfur molecule known as allicin, the primary flavor molecule associated with garlic when garlic tissue is cut or bruised. These amino acids are found only in allium vegetables like garlic, shallot, chive, and onion. Normally, both the enzyme and the cysteine sulfoxides are stored in separate compartments, but when the cells are damaged, these components intermingle and generate allicin. The more cells are damaged, the more flavor is released. Garlic that has been sliced will produce fewer allicin molecules than garlic that has been minced. The maximum amount of garlic flavor is produced when a clove is pulverized using a mortar and pestle, which crushes nearly all the cells.

KITCHEN TAKEAWAY The strength of garlic flavor released is proportional to how much it is cut or mashed, so grating, mincing, or crushing garlic will produce more flavor molecules than slicing it.

THE MAXIMUM AMOUNT
OF GARLIC FLAVOR IS
PRODUCED WHEN A
CLOVE IS PULVERIZED

WHY DOES THE FLAVOR OF GARLIC, BUT NOT THE FLAVOR OF CHILES, MELLOW WHEN COOKED?

THE SCIENCE The molecule primarily responsible for the flavor of garlic is allicin, a sulfur molecule generated when garlic tissues are broken or crushed. Allicin is also responsible for the sharp bite of fresh garlic. During cooking, two chemical processes happen that muzzle that pungency. First, heating garlic above 113°F/45°C causes the alliinase enzyme to irreversibly deactivate. Second, allicin is unstable and decomposes into hundreds of sulfur molecules when heated above 175°F/~80°C, which is well below the temperatures typically used for cooking. These secondary flavor molecules lack the "bite" of allicin and give cooked garlic its mellow flavor.

The hot pungency of chiles comes from capsaicin, a molecule that accumulates in the chile plant as it grows, rather than forming when the tissue of the vegetable is damaged, as is the case with allicin in garlic. Capsaicin is more thermally durable than allicin, remaining stable even at high temperatures, and it can dissolve out of a pepper from its skin and transmit its heat to a dish even when the pepper is left whole. Since drying happens at temperatures well below the decomposition point of capsaicin, dried chiles contain high concentrations of capsaicin and are also very effective at transmitting heat.

KITCHEN TAKEAWAY Garlic mellows during cooking because its key flavor molecule, allicin, is unstable and deactivates at a relatively low temperature. On the other hand, it takes a much higher temperature to break down the capsaicin in chiles, because capsaicin is a more stable molecule.

Do Beans Get Tough If You Add Salt or Tomatoes to Them at the Beginning of Cooking?

THE ANSWER No and yes

THE SCIENCE According to J. Kenji López-Alt, author of *The Food Lab*, salting beans doesn't make them tougher to cook. In fact, he found that salting beans actually improves their flavor and texture. What's the reason behind this? Magnesium and calcium ions are integral parts of beans' tough outer coating. When beans are soaked or cooked in salt water, the sodium ions begin migrating into the cells of the beans and replacing some of those magnesium and calcium ions. This process allows water to penetrate deeper into the bean and creates a more even distribution of moisture. The result? Beans that have been soaked

or cooked in salt water cook more evenly and yield a more tender, creamy result than those soaked or cooked in plain water. Beans soaked or cooked in plain water can end up rupturing during cooking, as the hot water softens the outer covering of the beans, then breaches it.

Cooking dried beans with tomatoes produces the opposite effect. Tomatoes contain a lot of citric acid. Citric acid reacts with calcium and magnesium, binding tightly to them and forming what is known as a citrate precipitate, which is extremely difficult to dissolve in water. So, when dried beans are cooked with tomatoes, the citric acid makes the beans' exteriors even more impervious to water, which keeps them from becoming tender.

KITCHEN TAKEAWAY Go ahead and salt your dried beans. If you like to presoak your beans, add one tablespoon of salt to one quart of water, and soak the beans for 8 to 24 hours at room temperature before cooking; otherwise, simply salt the cooking water. Either way, the salt will help soften the exterior of the bean and give a more evenly tender result. On the other hand, if your recipe calls for tomatoes, be sure not to add them until the beans are fully cooked.

Why Do Kids and Some Adults Hate Raw Broccoli?

THE SCIENCE Broccoli belongs to the *Brassica* genus of vegetables, which also includes cabbage, cauliflower, and mustard. Many of these plants are biochemically related and contain common enzymes that are released when the vegetable tissues are broken by chewing, cutting, or tearing. These enzymes then act on sugar-like compounds called glucosinolates (also stored in the plant's tissues) to produce bitter-tasting compounds known as isothiocyanates, which are designed to repel insects and animals that chew on the plant. But this natural predator repellant only works when the broccoli is raw; heat deactivates the enzymes. Though most individuals experience isothiocyanates as somewhat bitter, those with hypersensitive bitter taste receptors have a much more extreme negative reaction to raw brassica vegetables. This includes children, who have sensitive bitter taste receptors to keep them from eating poisons; this sensitivity usually lessens as they grow into adolescence.

KITCHEN TAKEAWAY When your kids tell you they hate broccoli, it may be because it tastes outright nasty to them. Try cooking it to crisp-tender, which should deactivate the enzyme responsible for creating that bitter taste. And be careful not to overcook it, because that can lead to the formation of different bitter-tasting flavor molecules.

 ## WHAT'S THE DIFFERENCE BETWEEN WHITE AND GREEN ASPARAGUS?

THE SCIENCE What we know as asparagus are actually the young shoots of a fern. If the plant is allowed to mature, the stems become woody, bitter, and inedible. Green asparagus is cultivated in the presence of sunlight, which induces photosynthesis. White asparagus is cultivated by covering the young shoots with soil to prevent their exposure to sunlight. As a result, the shoots lose their chlorophyll pigments (and green color), which allows sweeter notes to come through, instead

of the grassy flavor we are familiar with. A third type of asparagus, purple asparagus, is a special variety developed in Italy that contains more sugar and less fiber than its green and white relatives. Purple asparagus has a sweeter, nuttier flavor compared with the grassy flavor of green asparagus and the mild, buttery flavor of white asparagus.

KITCHEN TAKEAWAY To remove the tough end of an asparagus spear, snap it at the lower third of its stem. White asparagus must be peeled before cooking to remove its bitter outer layer, whereas green and purple asparagus can both be enjoyed as they are or peeled for increased tenderness.

Is There Really a Difference Between Starting Potatoes in Cold Water Versus Adding Them to Boiling Water?

THE ANSWER Yes

THE SCIENCE Potatoes contain an enzyme called pectin methylesterase, which interacts with pectin, the main component of the potato's cell walls, to keep the cells firm even when the vegetable is cooked. This enzyme has an optimal temperature of 140°F/60°C, so when potatoes are placed in a pot of cold water on the stove, they are allowed to spend a good amount of time at this optimal temperature as the water heats up, resulting in the perfect tender-but-still-firm texture. On the other hand, when potatoes are dropped directly into boiling water, the enzyme instantly deactivates, resulting in a mushier potato. Starting potatoes in cold water also allows the heat to transfer slowly throughout the potato, helping the starches evenly gelatinize; adding potatoes to boiling water results in uneven heat distribution and texture.

KITCHEN TAKEAWAY Potatoes should be started in cold water for an evenly cooked, tender-but-not-mushy result.

109

DOES THE KIND OF POTATO I USE WHEN BAKING, MASHING, OR FRYING REALLY MAKE A DIFFERENCE?

THE ANSWER Yes

THE SCIENCE There are many different varieties and cultivars of potatoes, each with their own distinct biochemistry and composition that affect texture, flavor, and mouthfeel. From the simplest perspective, potato chunks will do one of two things when they are boiled—disintegrate or stay intact. This is dependent on the type of starch and the amount of moisture available in the potato. All types of potatoes fall into one of three categories: starchy, waxy, or all-purpose.

Starchy potatoes, which include russets, Idahos, and most yams and sweet potatoes, are low in moisture and high in starch. When heated, their starch cells swell and separate, giving these types of potatoes a dry, fluffy texture when baked, a fluffy creaminess when mashed, and a crisp flakiness when turned into French fries.

Waxy potatoes, which include Red Bliss, creamers, and fingerlings, are

high in sugar and moisture and low in starch. Their starch cells are more loosely packed than those of starchy potatoes and don't separate when cooked; as a result, these potatoes retain their shape when boiled. When baked, waxy potatoes retain their firmness and don't become as fluffy as starchy potatoes do. They are not good candidates for frying, as their high moisture content will yield limp, soggy results, and they have a higher sugar content than other potatoes, which makes them brown faster.

All-purpose potatoes like Yukon Golds have a medium starch content, somewhere between starchy and waxy potatoes. These potatoes can be used for all types of cooking.

KITCHEN TAKEAWAY Starchy potatoes yield a fluffier, flakier texture, making them ideal for French fries, baked potatoes, and mashed potatoes, if you like yours light and fluffy. If you're looking for a potato that will hold its shape for potato salad and prefer smashed to mashed potatoes, waxy potatoes are the way to go.

WHEN BAKED, WAXY POTATOES RETAIN THEIR FIRMNESS AND DON'T BECOME AS FLUFFY AS STARCHY POTATOES

What Makes Popcorn Pop?

THE SCIENCE Corn is the fruit of the plant *Zea mays* and contains a hard kernel exterior that protects a dense, starchy interior. Popcorn is a specific variety of ancient cultivated corn that has been bred to contain just the right ratio of hard and soft starch granules, resulting in its ability to pop when heated.

But how exactly do the kernels pop? When a dried popcorn kernel is heated, steam forms inside its starchy interior. The hardness of popcorn's kernel shell is important at this stage—that is what keeps the steam locked inside, where it becomes pressurized and causes the starch to soften and gelatinize. (To put things in perspective, popcorn kernels are strong enough to withstand pressures nine to ten times greater than atmospheric pressure and temperatures as high as 356°F/180°C.) Eventually, the outer hull of the kernel ruptures and the gelatinized starches explode out with the rapidly escaping steam, creating the light, airy foam that we know as popcorn.

Much agricultural work has gone into developing strains that produce better tasting popcorn and leave fewer unpopped kernels. Orville Redenbacher was originally an agricultural scientist who tested thousands of strains of popping corn

WHEN POPCORN KERNELS ARE HEATED, STEAM FORMS INSIDE THEIR STARCHY INTERIOR

hybrids for years before settling on and perfecting the hybrid that would eventually help him capture much of the popcorn market back in the 1970s.

KITCHEN TAKEAWAY Jessica Koslow, chef/owner of the restaurant Sqirl in Los Angeles, recommends a high ratio of oil to popcorn (½ cup of oil to ⅓ cup of popcorn kernels) for an extra crunchy exterior. Oils used for popcorn should have high smoke points (grapeseed, avocado, sunflower, canola, and corn oils are good choices), since the kernels need to reach at least 356°F/180°C before they can pop.

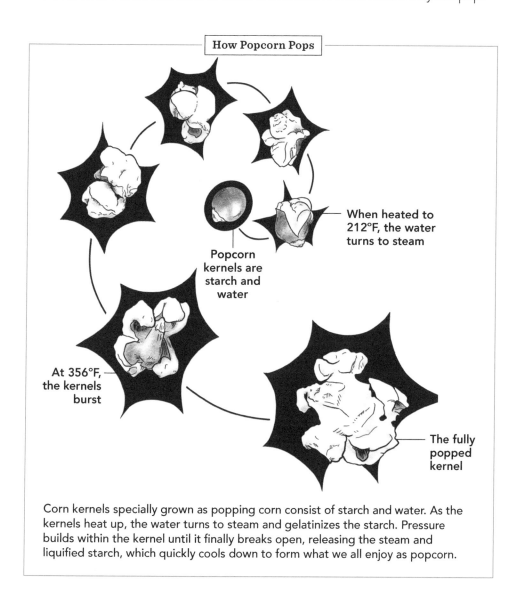

How Popcorn Pops

Popcorn kernels are starch and water

When heated to 212°F, the water turns to steam

At 356°F, the kernels burst

The fully popped kernel

Corn kernels specially grown as popping corn consist of starch and water. As the kernels heat up, the water turns to steam and gelatinizes the starch. Pressure builds within the kernel until it finally breaks open, releasing the steam and liquified starch, which quickly cools down to form what we all enjoy as popcorn.

Q WHAT GIVES MUSHROOMS THEIR MEATY FLAVOR?

THE SCIENCE While there are roughly 300 edible mushroom species on the planet, only 30 have been domesticated and 10 are grown on a commercial scale. The most widely produced and consumed mushrooms are button, shiitake, and oyster. All mushrooms contain guanylate, a flavor ribonucleotide that imparts savory umami, and its concentration is greater in mature and dried mushrooms. Some types of mushrooms contain higher guanylate levels, leading to stronger, more savory flavors.

Button or white mushrooms, which are sold as cremini mushrooms when slightly more mature and baby bella and portabella mushrooms when large and fully mature, are of the species *Agaricus bisporus*. White button mushrooms have a soft, creamy texture that becomes firmer and more flavorful as they lose water and mature into cremini and portobella mushrooms.

Shiitake mushrooms have a meaty texture similar to portobella mushrooms. When cooked, they have a smoky, earthy flavor that's partly attributed to the sulfur compound lenthionine. Dried shiitakes have among the highest guanylate levels of any known food, and they are an excellent way to boost the umami flavors of a dish. Unlike other mushrooms, shiitake mushrooms contain the polysaccharide lentinan, which can cause a rare allergic reaction called flagellate dermatitis in some individuals if the mushroom is eaten raw or undercooked; when the mushroom is fully cooked, lentinan degrades and causes no effect.

Oyster mushrooms (*Pleurotus ostreatus*) have a tender and tasty stem with a velvety cap. These mushrooms are prized for the delicate texture they impart to a dish. They also produce the aroma compound benzaldehyde, which gives them the slight odor and taste of anise. Oyster mushrooms have only about one-fifteenth the level of guanylates that shiitake mushrooms have.

KITCHEN TAKEAWAY To get the benefit of the mushroom's umami-producing guanylate without actually adding mushrooms to a dish, buzz dried shiitakes in a food processor until you have a fine powder, then stir this powder into soups, stews, or sauces; start with 1 teaspoon, then add more to taste.

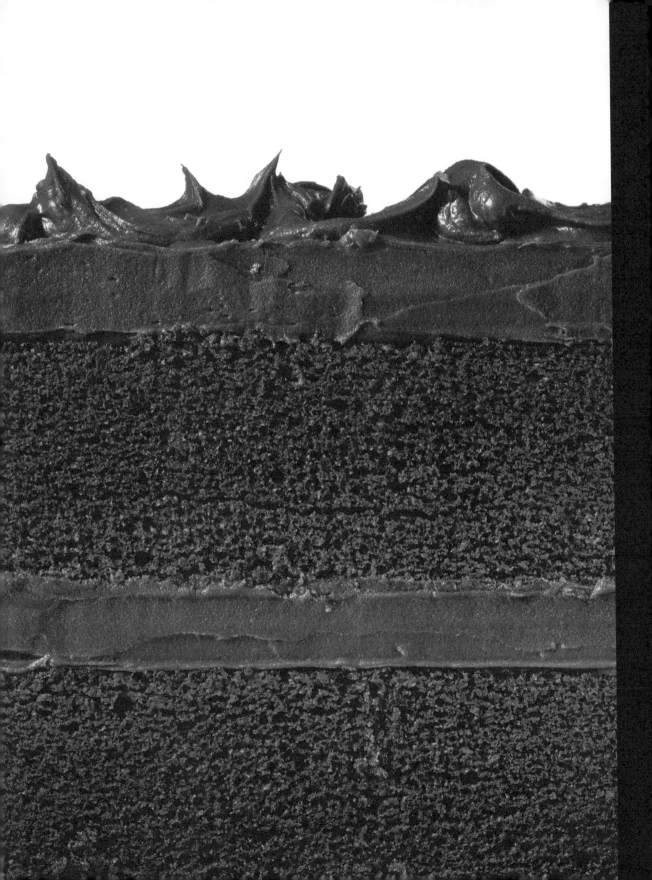

CHAPTER 6

Baking and Sweets

BAKING CAN BE CHALLENGING, ESPECIALLY WHEN YOU'RE FIRST STARTING OUT. Any slight changes to a recipe can result in a brick-hard loaf or a flattened disc instead of a tender cake. A little less baking soda or a little more sugar can throw off the balance of chemical reactions that need to happen during baking, leading to unappealing textures in the finished product. In this chapter, we'll explore some of the chemistry behind baking so you can troubleshoot your way to sweet success.

What Is Gluten?

THE SCIENCE Gluten is a stretchy network that forms when two proteins, glutenin and gliadin (found in wheat, rye, spelt, and several other grains) are mixed with water. When these proteins are hydrated, they immediately join together with what are known as disulfide bonds, forming protein strands. Then, protein-degrading enzymes called proteases go to work cutting up some of the protein strands, which then rearrange into smaller linear protein chains. The smaller protein chains form more disulfide bonds with other protein strands, creating a more complex network. Glutenin contributes elasticity and strength to this network, while gliadin contributes viscosity to the dough. Vigorous mixing or kneading keeps this gluten development going until the dough is smooth and stretchy.

Gluten is a welcome development for some baked goods, like rustic yeast breads, where the elastic network will provide the robust structure needed to capture the carbon dioxide bubbles created by the yeast, which will ultimately result in a deliciously chewy texture and open crumb in the finished bread. For baked items with tender, delicate interiors, such as cupcakes and biscuits, gluten

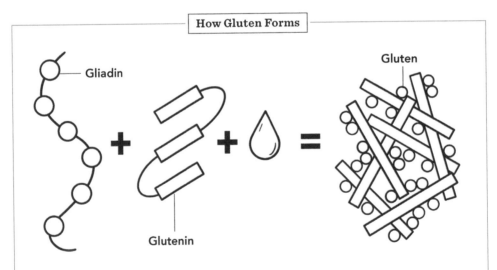

Gliadin and glutenin are proteins naturally found in wheat. When they are hydrated, they bond with one another to form gluten, a complex network of proteins chains.

development is minimized by mixing ingredients for a shorter amount of time, just until the batter or dough comes together.

Not all flours contain glutenin and gliadin, and those that do (like wheat flour) contain them in varying percentages. Wheat flour that contains a high percentage of glutenin and gliadin is sold as high-gluten flour or bread flour. All-purpose flour includes a lesser percentage of gluten-forming proteins, and cake flour contains even less.

KITCHEN TAKEAWAY Wheat flour (as well as several other grain flours) contains two proteins, glutenin and gliadin, which will form gluten when mixed together with water. Gluten networks provide structure to baked goods, and the strength of those networks can be controlled by the type of flour you use and how vigorously you mix together the batter or dough. For baked goods where tenderness is the goal, use all-purpose or cake flour (both of which have lower percentages of gluten-forming proteins) and keep mixing to a minimum; too much gluten development will make your cake or biscuits tough. For yeast breads where you want a crisp, chewy crust and airy interior, use high-gluten or bread flour and knead or mix the dough thoroughly.

Q HOW DOES YEAST WORK IN BAKING?

THE SCIENCE The yeast most commonly used in baking, *Saccharomyces cerevisiae*, is a type of fungus. When flour, water, and yeast are combined, amylase, an enzyme present in flour, breaks down the starches into sugars (maltose and glucose), which the yeasts start feeding on, producing alcohol (ethanol) and carbon dioxide (CO_2) as waste products. The carbon dioxide fills the little bubbles that are created in the dough during mixing. The elasticity of the gluten network (see What Is Gluten?, page 118) stabilizes those bubbles as they expand; as a result, the dough will rise as the yeasts continue to feast on the simple sugars and create more CO_2.

During this process, the carbon dioxide will also dissolve into the water in the dough to form carbonic acid, much in the same way that soda water is carbonated by dissolving carbon dioxide gas into water. When the dough is baked, the carbonic acid decomposes and releases carbon dioxide gas; at the same time, the water in the dough is converted to steam, and the yeasts, in the warmth of the oven, engage in one last feeding frenzy. The result is what is known as oven spring—the dough rises rapidly in the oven until the yeasts are finally killed by the rising temperature and the gluten network solidifies.

KITCHEN TAKEAWAY Yeasts are living organisms and will die if exposed to temperatures that are too high, above 140°F/49°C. Cold temperatures can make them sluggish. To check if your yeast still has life, mix it with a little lukewarm (110°F/~43°C) water and a bit of sugar to jump-start its activity. After about 5 minutes, you should see the mixture start to foam. If you don't see any activity, it's best to start with new yeast.

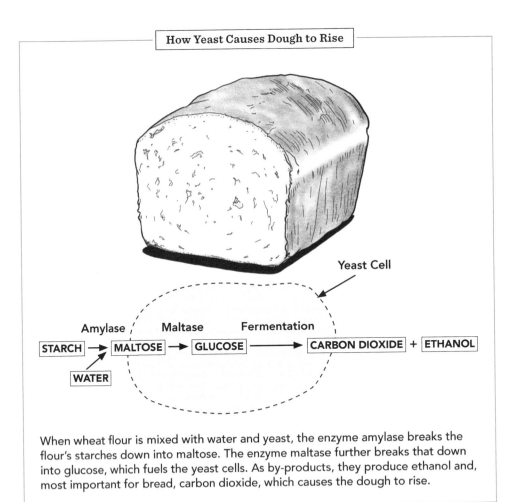

How Yeast Causes Dough to Rise

Yeast Cell

Amylase   Maltase   Fermentation

STARCH → MALTOSE → GLUCOSE ⟶ CARBON DIOXIDE + ETHANOL

WATER

When wheat flour is mixed with water and yeast, the enzyme amylase breaks the flour's starches down into maltose. The enzyme maltase further breaks that down into glucose, which fuels the yeast cells. As by-products, they produce ethanol and, most important for bread, carbon dioxide, which causes the dough to rise.

Why Are Some Breads Soft and Others Chewy?

THE SCIENCE Whether a bread is soft or chewy is strongly dependent on the amount of gluten protein found in the flour. Chewiness in a bread is a property of the elastic gluten network formed when flour is mixed with water; the chewier the bread, the more fully developed the gluten network is in the bread. Flour containing a higher percentage of gluten-forming proteins (such as bread flour) is used for rustic country loaves or baguettes, and all-purpose flour (which contains a lesser percentage of these proteins) is preferable for a soft sandwich bread, where you want structure and also a tender crumb and crust. While hydration is all that is needed to start forming gluten networks, vigorous mixing promotes the process, incorporating oxygen into the gluten network, which accelerates the unraveling of glutenin and gliadin proteins and the formation of disulfide bonds with neighboring proteins.

Other factors can affect the formation of the gluten network, like the presence of salt, which helps gliadin and glutenin entwine and produce a stronger network. Conversely, the addition of sugar and/or fat will disrupt the formation of the gluten network; sugar binds to water and removes some of the moisture necessary for gluten formation, and fat sheaths the proteins and prevents them from bonding. When you add sugar and/or fat to a yeasted dough, the result is a softer, richer bread like brioche or challah.

KITCHEN TAKEAWAY The factors that influence the chewiness of a bread are protein content in the flour and the presence of salt, sugar, and/or fat. Changing the ratios of these ingredients can help you change how your bread comes out.

CHEWINESS IN A BREAD IS A PROPERTY OF THE ELASTIC GLUTEN NETWORK

WHAT IS THE SCIENCE BEHIND NO-KNEAD BREAD?

THE SCIENCE Jim Lahey, the owner of Sullivan Street Bakery in New York City, developed a no-knead bread technique that was popularized in 2006 when Mark Bittman shared his excitement about the bread in the *New York Times*. In its simplest form, no-knead bread is made by combining flour, yeast, water, and salt in a bowl, letting the dough sit overnight, and placing it in a superheated Dutch oven to bake. The result is a crisp, crusty bread with much less mixing than is normally needed to properly develop the gluten in the dough. Instead, no-knead bread relies on hydration, enzymes, and time to produce a strong gluten network.

The initial mixing of the dough kick-starts both gluten development and fermentation. Left to rise for 12 to 18 hours, the dough is a hotbed of activity, with the proteins glutenin and gliadin bonding into gluten, then the protein-degrading enzymes called proteases chewing these gluten strands into smaller fragments that then recombine, strengthening the gluten network. The yeasts are also busy eating the sugars in the flour and generating carbon dioxide as a by-product. The bubbling of the carbon dioxide creates just enough movement in the dough to "micro-knead" the dough.

KEY TO THE SUCCESS OF NO-KNEAD BREAD IS THAT IT IS BAKED IN A COVERED SUPERHEATED CAST-IRON DUTCH OVEN IN A HOT OVEN

Key to the success of no-knead bread is that it is baked in a covered preheated cast-iron Dutch oven in a hot oven. Cast iron has the capacity to hold a tremendous amount of heat. When the dough is placed in the preheated Dutch oven and the pot is covered, the small space traps the moisture and creates a high-humidity environment, which is crucial to developing a delicious, light crumb and chewy outer crust. High humidity delays the formation of the crust, which allows the gases present in the dough to expand for longer as they heat up, creating the open structure we love so much in rustic breads. While this expansion of the dough is happening, the starches present on the surface of the dough start to absorb water and gelatinize, ultimately solidifying into a crackling, chewy crust.

KITCHEN TAKEAWAY Give it a try! Mix together 3½ cups of all-purpose flour, ½ teaspoon of instant yeast, 1½ teaspoons of salt, and 1¾ cups of water to form a dough. Cover with plastic wrap and let the dough rest at room temperature for 12 to 18 hours. Transfer the dough to a work surface generously dusted with flour. Fold it several times just until the dough it no longer sticky. Shape it into a round loaf, with the seam underneath it. Cover with plastic wrap and let it rest for an hour. After 30 minutes, put a large covered Dutch oven in a 450°F/~230°C oven and preheat it for 30 minutes. Take the Dutch oven out of the oven, remove the lid, and carefully place the dough inside, seam side up (watch your hands). Cover the pot immediately, and bake for 30 minutes. Remove the lid and bake until the top of the bread has formed a nice brown crust, another 20 to 30 minutes. Carefully remove the bread from the pot (use tongs) and let cool on a wire rack before slicing.

WHY DOES BREAD HARDEN WHEN IT GETS STALE BUT COOKIES AND CRACKERS GET SOFT?

THE SCIENCE The starches in the flour used to make baked goods are normally trapped in small, hard granules. When flour is hydrated into a dough and baked, those granules burst and release starch that has expanded from absorbing moisture.

When baked goods age, moisture moves around in an attempt to establish equilibrium between the interior and exterior, migrating from high-moisture regions to low-moisture regions. In the case of bread, the water present in the starches inside the bread moves toward the drier crust and evaporates out into the environment. As these starches lose water, their primary components, the polysaccharides amylose and amylopectin, realign into their original compact structure that creates a hard texture and expels even more moisture out of the bread. The result—a hard, stale loaf of bread.

The starches in cookies and crackers go through the same process—moisture shifts from areas of high to low concentration and the starches harden. However, cookies contain a large amount of sugar and crackers contain a large amount of salt. Salt and sugar are both hygroscopic, meaning they are adept at absorbing moisture from the air. As a result, more moisture is absorbed by cookies and crackers than is expelled through evaporation, and that increased moisture offsets the effect of the starches hardening. The net result is softened cookies and crackers.

KITCHEN TAKEAWAY Resuscitate stale cookies by putting them in a plastic bag with a slice of bread. The moisture will slowly transfer from the cookies to the bread, and the cookies will regain their crispness.

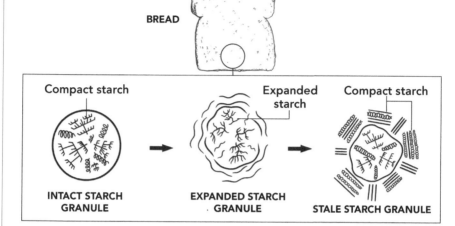

BREAD

Compact starch

Expanded starch

Compact starch

INTACT STARCH GRANULE

EXPANDED STARCH GRANULE

STALE STARCH GRANULE

The starches in raw flour are normally tightly packed inside granules. When baked into bread, crackers, or cookies, these granules break open and the starch molecules expand in size as they absorb water from the dough. During the process of staling, the starches release this water; in bread, the water evaporates through the crust, the starches resume their original compact form, and the bread hardens. Because crackers and cookies contain significantly more sugar and/or salt than bread, both of which attract water, this effect is countered. Additional moisture is absorbed from the air and the result is softening.

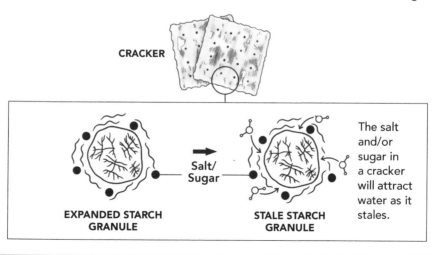

CRACKER

Salt/ Sugar

EXPANDED STARCH GRANULE

STALE STARCH GRANULE

The salt and/or sugar in a cracker will attract water as it stales.

What's the Difference Between Baking Soda and Baking Powder?

THE SCIENCE Baking soda, also known as sodium bicarbonate, is an alkali that releases carbon dioxide when it's heated above 175°F/80°C and/or mixed with an acid, helping doughs and batters puff up in the oven. Though heat alone is enough to trigger some reaction, baking soda's leavening properties are truly activated when the alkali is combined with an acidic ingredient like buttermilk or yogurt. Plus, the addition of acid neutralizes baking soda's metallic taste. The biggest challenge with baking soda is that it is fast-acting, so it's best used in thick batters and doughs that can trap or weigh down the carbon dioxide bubbles to prevent a loss of structure.

Because of baking soda's quick-acting nature, a more stable alternative was sought, which is how baking powder was developed. Baking powder is made by mixing sodium bicarbonate (baking soda) with a solid acid—ammonium sulfate, alum (sodium aluminum sulfate), cream of tartar (tartaric acid), or bone ash (monocalcium phosphate). It is formulated so that the sodium bicarbonate first reacts with the acid when hydrated in the batter or dough, then reacts with the acid again when heated, releasing carbon dioxide bubbles each time to lighten the dough and help it rise in the oven. Baking powder has only one-third to one-fourth of the leavening strength of baking soda by weight. Sometimes a recipe will call for both baking soda and baking powder for extra leavening power; this is because baking soda increases the alkalinity of the bake as it decomposes into sodium carbonate when heated, speeding up the very desirable Maillard reactions that contribute to browning and flavor (see Why Do Foods Brown? The Maillard Reaction, page 12).

KITCHEN TAKEAWAY Baking soda can be substituted for baking powder as long as there's a source of acid included in the batter (yogurt, buttermilk, lemon juice, or cream of tartar, for example). Baking soda has three to four times the leavening power of baking powder, so if you are swapping in baking soda, you should use ⅓ to ¼ the amount of baking powder called for in a recipe. Be careful, though, because too much baking soda can impart a chemical taste.

DO I NEED TO USE CAKE FLOUR WHEN I BAKE A CAKE?

THE ANSWER You should if the recipe asks for it

THE SCIENCE Cake flour is finely milled from a soft wheat, meaning that it has a lower gluten protein content than other flours, around 7 to 9 percent. In contrast, bread flour has a protein content of 12 to 15 percent, and all-purpose flour 10 to 12 percent. Normally, the formation of gluten is desirable, as it helps entrap the carbon dioxide released by leavening agents, allowing the baked good to rise, and it produces chewy textures. But for cakes, the goal is a light, tender texture, which means minimal gluten development is key.

The fact that cake flour is bleached with chlorine also aids in the achievement of a moist result. Chlorinating gluten proteins improves their solubility, which helps to create a more cohesive batter. Chlorine also breaks down the starches into smaller fragments and adds a chlorine atom to the starch structure, which increases the starches' ability to hold onto water and bind to fats. This enhances the even dispersion of batter ingredients, resulting in a moister cake.

KITCHEN TAKEAWAY While it's important to use the right flour for the best possible results, if you don't have cake flour on hand, you can substitute 1 cup of cake flour with ¾ cup of bleached all-purpose flour and 2 tablespoons of cornstarch to reduce the overall amount of protein in the batter.

CHLORINATING GLUTEN PROTEINS
IMPROVES THEIR SOLUBILITY,
WHICH HELPS TO CREATE
A MORE COHESIVE BATTER

Does the Temperature of My Butter Matter When I Use It in a Recipe?

THE ANSWER Yes

THE SCIENCE Butter is made of water, milk proteins, sugars, and butterfat. Butterfat is the predominant component and is composed of a variety of different fats with different melting points. Butter that's been chilled in the freezer contains ice crystals and crystalline fats. Chilled butter retains its shape until it is warmed to room temperature, at which point some of the fats have melted while others remain solid and crystalline. Softened butter contains enough melted fats and liquid water that the crystalline fats can begin sliding across one another. When softened butter is creamed with sugar, air becomes incorporated into the solid butterfat structure and forms air pockets. These air bubbles expand when heated and contribute to the fluffiness of the final baked product.

On the other hand, when butter is melted, all the fats become completely liquid. Melted butter behaves more like oil and does not hold structure in the same way that softened or chilled butter does. When air gets whipped into melted butter, the melted butter is too weak to trap the air. This leads to a denser bake, so for those who love a dense brownie, melted butter is the way to go.

KITCHEN TAKEAWAY Whether you melt, soften, or chill butter depends on the final texture you want in your baked good. Melted butter gives baked goods a dense, fudgy texture that's great for brownies or pudding cakes. Softening butter (letting it come to room temperature) allows you to cream sugar into it and create air pockets that will yield cakes with tender, light crumbs. Chilled butter keeps its solidified structure during baking, which helps create flakiness in pie crusts and croissants.

Q DOES IT MATTER HOW YOU MIX A CAKE BATTER?

THE ANSWER Yes

THE SCIENCE When you bake a cake, you're relying on some very intricate science and chemistry to create a delicious, sugary treat. One of the most important features of a cake is its soft texture and crumb. Much of a cake's texture derives from the gas bubbles that are incorporated into the batter during the mixing process. Creaming butter and sugar together traps air bubbles in the batter. These air bubbles are important because they function as leaveners would, providing lightness to the baked good. For that reason, it's important that you introduce as much air into the mixture as possible. In recipes that instruct you to cream butter and sugar together, you'll often see the direction to beat until the mixture lightens in color and becomes fluffy, indicators that a sufficient amount of air has been whipped in.

After the butter and sugar have been creamed together, it's time to beat in the eggs. Beating denatures (breaks down) the egg proteins, which allows them to form a thin layer around the air bubbles. Sheathing the air bubbles with fat stabilizes them, keeping them from collapsing and resulting in a dense, deflated finished product.

Next, you add the flour. When wheat flour is incorporated, care must be taken to minimize gluten development; this is why many cake recipes warn you not to overmix the batter at this stage. Some gluten is necessary to help support the cake's structure, but too much gluten will result in a tough, not tender, cake. Stir the flour in just enough so that you no longer see any white streaks.

KITCHEN TAKEAWAY For the best results, follow your recipe directions closely, especially when it comes to how thoroughly your dough or batter should be mixed.

CREAMING BUTTER AND SUGAR TOGETHER
TRAPS AIR BUBBLES IN THE BATTER

Does It Matter if You Use a Glass or Metal Pan When Baking?

THE ANSWER Yes

THE SCIENCE When you put your baking dish or pie plate into a preheated oven, heat begins to transfer from the oven to the sides and bottom of the pan, then to the dough or batter. Glass, which is a heat insulator, takes more time to heat up and cool down than metal, which is a better conductor of heat. As a result, it will take longer for a glass pan to heat up and transfer that heat to its contents. The advantage of this is that glass pans distribute heat better than metal pans do, which promotes even baking, and they will stay warm longer after they are taken out of the oven. Glass pans are also unreactive to eggs and impervious to acidic foods like tomatoes and citrus.

Metal pans can withstand higher temperatures than glass pans can, and they are also safe to use under the broiler (glass pans can shatter at high temperatures). Since metal conducts heat faster than glass does, foods baked in metal pans will brown faster and more deeply. Light metal pans, like aluminum, will reflect heat and thermal infrared light (one of the ways heat is emitted by the oven heating coils and reflected by the walls of the oven), resulting in food browning more slowly than if it were baked in a dark metal pan, which absorbs infrared light. A light metal pan will still heat up faster and brown food more deeply than a glass pan will.

KITCHEN TAKEAWAY If using a glass pan, lower the oven temperature by 25°F/14°C and add 10 minutes to the baking time to account for the differences in heat transfer. Glass may not heat up quickly, but once it has heated up, it stays hot and can overbrown food near the end of cooking or baking if kept at the same temperature as a metal pan. Also, use dark metal pans when baking items with short bake times to promote browning.

DOES IT MATTER WHERE I BAKE
MY CAKE IN THE OVEN?

THE ANSWER Yes

THE SCIENCE Ovens are essentially insulated metal boxes with two heating elements, one placed on top and the other on the bottom. When an oven is set to preheat, these elements turn on and increase the temperature of the oven. Once the oven has reached the desired temperature, a thermostat registers the temperature and turns off the electricity to the elements. However, when the oven minutely cools down by radiating heat out of the oven's sides, the thermostat kicks the bottom heating element back on to bring the oven back to the desired temperature. The thermostat tends to be situated closer to the bottom heating element in most ovens, so the temperature is more tightly regulated at the bottom of the oven. The trouble is that heat rises, so the top of the oven can end up getting much hotter than the preset temperature. Because of this unequal heat distribution, foods placed on the top rack will cook faster and brown more overall than those placed on the bottom rack. On the other hand, foods placed on the bottom rack will brown faster on the bottom, because they are closer to the radiating heat of the bottom heating element. Convection ovens can even things out by blowing the hot air around the oven and redistributing the heat.

KITCHEN TAKEAWAY If you don't have a convection oven and have several pans in the oven at the same time, be sure to rotate them once or twice during the baking time to help them bake evenly. The middle rack is the best place to cook food evenly. If you want to promote browning, place your food on the top or bottom rack of the oven.

Q WHAT CAUSES BAKED GOODS TO PUFF UP IN THE OVEN AS THEY BAKE?

THE SCIENCE Baked goods require moisture and/or the work of a leavening agent, like yeast, baking powder, or baking soda, to puff up during baking. When heated, baking soda and baking powder break down and release carbon dioxide, forming bubbles in the batter or dough. As the batter or dough continues to bake and eventually reaches a boiling point, water present in the batter or dough will convert to steam and expand the carbon dioxide bubbles further, causing the cake, bread, or cookies to rise. The starches in the batter or dough will start to gelatinize and ultimately solidify around these bubbles, in the case of cake or bread, creating its crumb.

Yeast breads rise in the oven in a somewhat similar fashion. In the first 10 to 20 minutes of baking, yeasted bread doughs reach a point called oven spring, where expanding steam and a final flurry of yeast activity cause the dough to rise rapidly in the oven until the yeasts are finally killed by the rising temperature and the crust hardens, making further expansion impossible.

Some baked goods, such as puffed pastry, rely solely on moisture to create puffiness. In these recipes, the water in the dough converts to steam to push the pastry layers apart before evaporating, leaving behind buttery, airy crispness.

KITCHEN TAKEAWAY You can control how much your cakes, breads, and other treats puff up during baking by tinkering with the ratios of moisture, yeast, and/or baking soda in your batter or dough. But be careful—if your dough isn't strong enough, too much yeast and baking soda can cause your baked goods to collapse!

Why Are Some Cookies Chewy and Others Crispy?

THE SCIENCE The texture of a cookie is a function of its ingredients, specifically the type of flour, fat, sugar, and egg used (or not) in the recipe. If you're after a chewy cookie, use cake flour, vegetable shortening, brown sugar, and eggs. The lower protein content of cake flour results in reduced Maillard browning. Shortening has a high melting point, which keeps

the dough stiff for longer, minimizing spread during baking. Brown sugar contains moisture that helps keep the cookie moist as it heats up. Using honey or corn syrup as the sweetener also helps retain moisture, yielding a softer cookie. Egg contains moisture and binds together and coagulates the dough as it bakes, resulting in a softer and thicker cookie. Crispy cookies are best made using a high-protein flour, oil or butter, granulated sugar, and no egg. The high protein supports the Maillard reaction (see Why Do Foods Brown? The Maillard Reaction, page 12), which promotes browning. Granulated sugar contains little moisture compared with brown sugar, which yields a drier batter and helps with crisping. A cookie dough's ability to spread easily during baking contributes to crisping and this combination achieves that in two ways: leaving out eggs and using either oil or butter (butter because of its lower melting point than shortening).

KITCHEN TAKEAWAY Armed with the basics of cookie chemistry, you can tweak recipes with the flavors you love to create perfectly textured cookies every time. For crispier cookies, choose high-protein or all-purpose flour, oil or butter, and granulated sugar. For softer, moister cookies, go for cake flour, shortening, and corn syrup, honey, or brown sugar.

BROWN SUGAR CONTAINS MOISTURE
THAT HELPS KEEP THE COOKIE MOIST AS IT HEATS UP

WHAT IS THE SECRET TO A FLAKY PIE CRUST?

THE SCIENCE The foundation of a pie crust is flour, fat, and water. Whether the fat is butter, lard, or vegetable shortening, it's mixed or cut into the flour until the mixture is crumbly and the fat is broken down into pea-size pieces. This process coats the gluten-forming proteins gliadin and glutenin with fat, which will impede the formation of gluten once the water is added to the dough; the formation of too much gluten can result in a tough, not flaky, pie crust. Next, when you add water to your flour-and-fat mixture, it is important that you minimize your manipulation of the dough (and therefore minimize the formation of gluten), mixing it just until it comes together, then rolling it out and fitting it into the pie plate.

Before filling and baking the pie dough, let it chill completely in the fridge or freezer; you want the fat in the dough to be solid when it goes in the oven. It's also important the oven is properly preheated before you bake the pie. In a hot oven, the water in the dough will quickly vaporize into steam, which will push the dough apart into flakes. If the fat isn't cold enough and/or the oven is not hot enough, the fat will melt and be absorbed by the dough before it has time to set, resulting in a soggy crust.

KITCHEN TAKEAWAY One trick to creating a flakier pie crust is to substitute some of the water with vodka. Alcohol slows down the formation of gluten, giving the crust a more tender texture. The vodka evaporates quickly, leaving behind no off-flavor.

Q IS THERE ANY DIFFERENCE BETWEEN TYPES OF SUGARS?

THE ANSWER Kind of

THE SCIENCE The common form of table sugar is also known as sucrose, a sugar molecule made up of the smaller sugar molecules glucose and fructose, known as simple sugars. Most of the world's table sugar is extracted from sugarcane or sugar beets. To produce granulated and crystallized sugars, the sugar-rich plant material is sliced or crushed and soaked in water to dissolve the sugars, then squeezed to extract the sugar syrup. The sugar syrup is boiled until the concentration of sugar is high enough that the sugar crystallizes as the syrup cools down. The raw crystallized sugars are then separated from the liquid by centrifugation.

The raw sugar is carefully melted down, filtered with activated charcoal to remove impurities and off-colors, and recrystallized a second time to obtain pure crystalline sugar, or what we know as granulated white sugar.

Sugar is naturally white, but the boiling process causes the sugars and impurities in the raw syrup to caramelize and break down into a dark liquid that eventually becomes molasses. Molasses still contains 75 percent sugar, but it's not economical to further crystallize the product. Brown sugars and what is marketed as "raw" sugar are sugars that still contain some molasses or, more commonly, are refined white sugar that has had molasses added back in to give it its brown color.

KITCHEN TAKEAWAY Brown and raw sugars have lightly caramelized flavors due to the presence of molasses, which has either been left in during the refining process or added to fully refined white sugar.

Why Does Honey Crystallize?

THE SCIENCE Bees produce honey by gathering nectar from flowers. Nectar itself is a sweet solution with an average concentration of 23 percent simple sugars, including glucose, fructose, and sucrose. When bees absorb nectar, they release salivary enzymes that convert each molecule of sucrose into one molecule of glucose and one molecule of fructose, increasing the overall concentration of sugars. The predigested nectar is regurgitated in the hive, where other bees repeatedly ingest and regurgitate the nectar until its partially digested. This process creates air bubbles in the nectar that increase its surface area and causes some of the water to evaporate. The partially digested nectar is moved to storage honeycombs, where more water is evaporated by the heat of the hive and the fluttering of bee wings, which recirculates air around the hive. The sugar content of the nectar continues to rise as more moisture is lost, until it reaches 75 percent and honey is formed. Honey is about 75 percent glucose and fructose and 20 percent water, with the remainder composed of acids and enzymes. At this concentration, the solution is supersaturated, which means the amount of sugars dissolved in the honey is much higher than it normally would be at room temperature. The highest concentration of glucose that can fully dissolve in water is about 48 percent. On the other hand, fructose is extremely soluble in water and can form an 80 percent solution. So if the glucose concentration in a jar of honey is higher than 48 percent, the glucose molecules will tend to fall out of solution and crystallize.

KITCHEN TAKEAWAY There's nothing wrong with honey that has crystallized, and it can be re-dissolved by heating it over a hot pot of water or microwaving it in 30-second increments until all the sugar has dissolved.

WHAT MAKES PEANUT BRITTLE BRITTLE?

THE SCIENCE Peanut brittle has an amazing snap and crisp texture that makes you want to grab more and more. The secret ingredient that makes that cracking texture possible isn't sugar or butter—it's baking soda. The first step of making peanut brittle is to boil water, granulated sugar (sucrose), and corn syrup together until the mixture reaches 300°F/150°C, known as the hard crack stage. At this point, most of the water has evaporated and the solution is supersaturated with sugar; it is also super unstable and prone to crystallize (which would result in a granular texture) were it not for the presence of the corn syrup. Commercial corn syrup contains acid, which helps the sucrose react with the water to form glucose and fructose. This is important because fructose is far more soluble in water than sucrose is. Also, the presence of all three sugar compounds disrupts the formation of crystals by preventing any single type of sugar from becoming dominant when the solution becomes more concentrated. Mixtures of sugars have a harder time crystallizing because the different molecules constantly bump into one another and destroy one another's crystal structures.

You can test for this stage by dripping a bit of the sugar syrup into cold water—it will separate into hard, brittle threads. Sugar cooked to the hard crack stage will remain in a non-crystalline form when it cools and hardens, yielding a smooth texture.

To finish the peanut brittle, the sugar syrup is taken off the heat, and peanuts, butter, and baking soda are stirred in. The high heat immediately decomposes the baking soda into carbon dioxide gas and sodium carbonate. The carbon dioxide creates air pockets in the cooling syrup, which give the brittle its signature snap. The proteins in the butter and peanuts react with the sugars to form Maillard flavor compounds, amplified by the presence of the baking soda, which promotes browning, yielding the brittle's caramel flavor and color.

KITCHEN TAKEAWAY If you're having a hard time getting your peanut brittle to the hard crack stage, leaving you with something chewy rather than crispy, bring the mixture to 302°F/151°C to drive off more moisture. Sometimes, the surrounding humidity can affect the temperature at which the sugars reach the hard-crack stage.

What Does It Mean to Temper Chocolate?

THE SCIENCE When something is tempered (whether it be chocolate or iron, for instance), it is slowly heated to allow its molecules to rearrange themselves in such a way that the quality of the material is improved and strengthened. The cocoa butter in untempered chocolate is a mixture of different crystalline forms of fat in various sizes. When tempered, the cocoa butter yields a firmer chocolate with a glossy finish. Chocolate is tempered by first heating it to between 110°F and 120°F/43° and 49°C to liquify all the cocoa butter crystals. The melted chocolate is then slowly cooled to 82°F/28°C; doing so allows the cocoa butter to form a specific crystal, called form-5 beta crystals. This stable fat crystal is what gives well-tempered chocolate its impressive glossy sheen and sharp snap.

The tempering process can be aided by "seeding" melted chocolate with finely chopped tempered chocolate. The beta crystals in that chocolate serve as points around which new beta crystals can organize and grow.

KITCHEN TAKEAWAY As a home cook, you really only need to consider tempering if you want a high-quality finish on chocolate you're using for dipping, coating, or making molded candies. To temper chocolate, chop or grate your chocolate, and place ⅔ of the chocolate in the top of a double boiler set over simmering water. Carefully heat the chocolate to 110°F to 120°F/43°C to 49°C (use a candy thermometer to measure this), stirring until it's smooth. Remove the top pan from the double boiler and let the chocolate cool to 95°F to 100°F/35° to 38°C, then slowly stir in the remaining unmelted chocolate until the overall temperature drops to 82°F/28°C. Return the pan to the double boiler, heat the chocolate to 88°F/31°C, then pour it into molds or use it for dipping or coating.

WHY DO THE FLAVORS OF CHOCOLATE AND COFFEE GO TOGETHER SO WELL?

THE SCIENCE Chocolate is produced by harvesting fully ripe pods from the cacao tree, *Theobroma cacao*, removing the beans from the pod, and fermenting them for seven days to remove the pulpy fruit that encases them. The beans are then cleaned, dried, and roasted. Similarly, coffee beans are produced by picking the fruit of the *Coffea arabica* plant, called coffee cherries, and fermenting them for a day or two to remove the pulp. The coffee beans are removed and dried for several weeks, then cleaned and roasted.

The combination of the fermentation and roasting steps impart similar flavor profiles to coffee and chocolate. As microorganisms digest the fruity pulp starches, sugars and amino acids (leucine, threonine, phenylalanine, serine, glutamine, and tyrosine) are generated in both and serve as flavor and aroma precursors. When the beans are roasted above 284°F/140°C, the sugars react with those amino acids through Maillard browning to generate complex flavor compounds, mostly pyrazines and aldehydes, that are key components of the flavors we recognize as chocolate and coffee. When coffee is added to a chocolate cake batter or chocolate and coffee are combined together in other sweets, those shared flavor compounds deepen and enhance one another.

KITCHEN TAKEAWAY For more complex chocolate flavor in a cake or cookies, swap out two to four tablespoons of the recipe's liquid for strong cold-brewed espresso.

THE COMBINATION OF THE FERMENTATION AND ROASTING STEPS
IMPART SIMILAR FLAVOR PROFILES TO COFFEE AND CHOCOLATE

CHAPTER 7

Food Safety and Storage

YOU MAY WONDER FROM TIME TO TIME IF THE FOOD YOU'RE CONSUMING IS TRULY SAFE. When the baked fish on your plate comes from Chile and the sliced garlic comes from China, you might have concerns about picking up food-borne diseases from around the world, or you might wonder what governmental regulations are set in place to keep your food safe from contamination. Also, when you bring home groceries or spend the day prepping make-ahead meals, you might be unsure as to the safest way to store each type of food. Proper food handling and storage helps prevent or slow down the growth of mold, yeast, and bacteria, and learning a bit about the chemistry behind these processes can help keep you safe and healthy as you explore the world of cooking.

Should Hot Food Be Put Immediately in the Fridge or Left Out to Cool First?

THE SCIENCE There's a common belief that food should be kept out to cool down before placing it in the fridge. The rationale behind this is that putting warm food in the fridge will increase the temperature inside and put the other food at risk of being in the bacterial "danger zone." Today's refrigerators have a built-in thermostat that detects changes in temperature and will quickly accommodate the influx of heat from hot foods. The amount of air circulating through the refrigerator will also reduce the likelihood that the hot food will affect other foods around it. In comparison, by leaving your warm food out to cool to room temperature, bacteria can quickly multiply in the range of 40°F to 140°F (5°C to 60°C) and double their population every 20 minutes at room temperature, so bacteria populations can become quite large in just a few hours.

KITCHEN TAKEAWAY It's best to place hot food being prepared in advance immediately in the fridge to avoid possible bacteria growth. You can separate it into smaller containers to help it cool down faster.

IS RAW MILK SAFE TO DRINK?

THE ANSWER No

THE SCIENCE Raw milk can contain bacteria, such as *Staphylococcus aureus*, *Escherichia coli*, *Listeria monocytogenes*, and *Salmonella typhimurium*, that can make you very sick. During the pasteurization process, raw milk is heated to a minimum temperature of 161°F/72°C for 15 seconds, which kills at least 99.9999 percent of the pathogenic bacterial strains commonly found in milk, making it safe to drink and extending its refrigerator shelf life to two weeks. For shelf-stable milk that will keep for six to nine months unrefrigerated, another method, ultra-high-temperature pasteurization, is used; during this process, the raw milk is heated above 275°F/135°C for 2 to 5 seconds. (Though this seems like a great option, it's worth noting that ultra-high-temperature pasteurization does affect the taste and smell of the milk, since the higher temperatures cause Maillard browning.) Studies have shown that the nutritional differences between pasteurized and raw milk are few. The risk of bacterial infection is much higher with raw milk, and that risk is compounded if you drink raw milk regularly.

KITCHEN TAKEAWAY The minor differences in nutritive value between pasteurized and raw milk do not outweigh the higher potential for consumption of harmful bacteria from raw milk; the same goes for soft or soft-ripened unpasteurized cheeses like mozzarella, Brie, and queso fresco. On the other hand, unpasteurized raw milk cheeses that have been aged for at least 60 days are safe to eat, since their high salt, acid, and mold contents create an environment that's inhospitable to pathogenic bacteria.

Is It Safe to Eat a Raw Egg?

THE ANSWER Depends—how lucky do you feel?

THE SCIENCE Raw eggs are typically associated with *Salmonella* because chickens carrying the bacteria can pass it to the interior of the egg before the shell has formed. *Salmonella* can also attach to the eggshell through exposure to chicken droppings. Of course, not all raw eggs carry *Salmonella*; there's just no way to know whether a particular egg may have it. Raw eggs that have been pasteurized have much fewer active *Salmonella* bacteria and are safer to eat. Eggs that have been cracked, damaged, or contain dirt are more likely to harbor *Salmonella*. The presence of *Salmonella* on an egg may pose a greater risk for young children, older adults, and those with weakened immune systems.

KITCHEN TAKEAWAY If you want to reduce your possible exposure to *Salmonella* as much as possible, use pasteurized eggs for raw egg preparations like homemade mayonnaise or steak tartare, as well as dishes like hollandaise sauce, in which the eggs are heated below 160°F/71°C, the temperature that kills *Salmonella*. Note that washing the outside of the egg before using it will not remove *Salmonella* bacteria.

Q DO I HAVE TO WORRY ABOUT MY EGGS GOING BAD?

THE ANSWER No—if they're bad, you'll know it

THE SCIENCE As an egg ages, the shell becomes more porous, which means that moisture from inside the egg can begin to evaporate through the shell. Over time, an air pocket forms inside the egg. The egg yolk starts to absorb moisture from the egg white, becoming larger and easier to break. Proteins in the egg white denature as they absorb oxygen from the air pocket, and they will eventually lose their ability to thicken and foam into a meringue when whipped. Some spoilage microorganisms can also begin to grow in the egg as it ages, changing the color of the egg white to pink or iridescent. Generally speaking, commercial eggs that have been washed can last up to five weeks in the refrigerator. Unwashed freshly laid eggs, which are protected from outside bacteria by a thin secretion (called a cuticle) that coats the outside of the shell, can last up to three weeks on the counter and up to three months in the refrigerator.

KITCHEN TAKEAWAY The best judge of whether an egg is good or not is your nose. If it smells bad, it is bad. To test the freshness of eggs you plan to separate (to make whipped egg whites, for example), before you crack them open, put them in a bowl of water; if they float, they contain a large air pocket, meaning that the whites are over the hill and won't whip up properly.

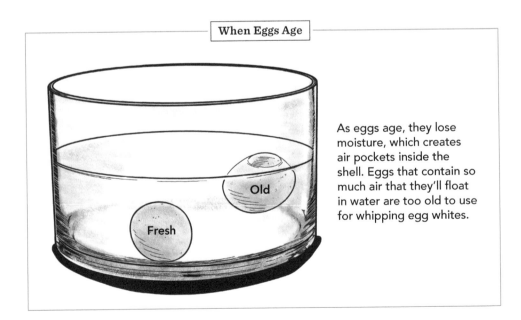

When Eggs Age

Old

Fresh

As eggs age, they lose moisture, which creates air pockets inside the shell. Eggs that contain so much air that they'll float in water are too old to use for whipping egg whites.

IS RAW COOKIE DOUGH
OKAY TO EAT?

THE ANSWER No

THE SCIENCE While it might be tempting to eat a spoonful of cookie dough, whether it is homemade or from your supermarket's refrigerator case, resist the urge! The raw eggs in home-made dough can contain *Salmonella*. Commercial cookie doughs circumvent that by using egg power that has been heat-treated to remove *Salmonella*, but both commercial and homemade cookie doughs contain uncooked flour, which can still harbor heat-resistant strains of *Salmonella*, *Escherichia coli*, and other bacteria.

KITCHEN TAKEAWAY The dough in commercial cookie dough ice cream is specifically heat treated to make sure that it's safe to eat, whereas packaged and homemade raw cookie doughs are not. Homemade doughs should be baked into cookies before consuming.

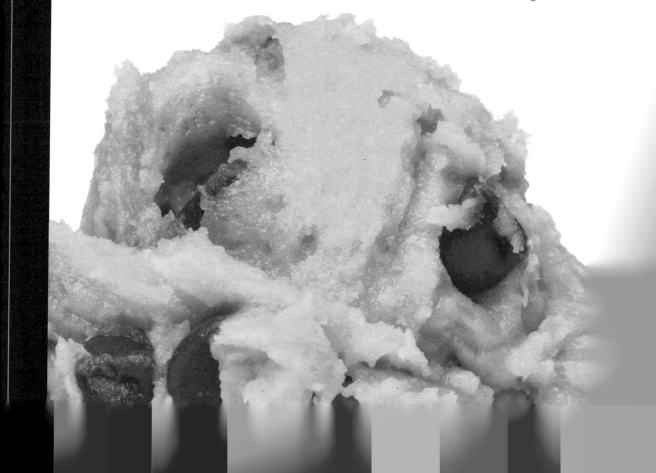

Should Raw Chicken and Turkey Be Washed Before Cooking?

THE ANSWER No

THE SCIENCE The surfaces of poultry are potential breeding grounds for bacteria that can make you sick, like *Campylobacter* and *Salmonella*. Washing poultry can lead to cross-contamination with other foods as contaminated water splashes around in the sink. And without scrupulous cleanup afterward, chances are that some bacteria will remain and colonize in the very surface you use to clean your food and dishes. Brining has been shown to slow down the growth of pathogens on chicken and turkey, but if you rinse the brine off, the pathogen numbers can rise again.

KITCHEN TAKEAWAY Chicken and turkey should travel straight from their packaging to the cutting board, if necessary, and then to the skillet or roasting pan. Rinsing increases the chances of bacterial contamination.

IS IT OKAY TO EAT UNDERCOOKED PORK?

THE ANSWER No

THE SCIENCE Pigs can become infected with *Trichinella spiralis*. Pigs get infected by this parasite when they eat garbage meat scraps that contain the roundworm's larvae. Infected swine do not show many clinical symptoms of infection, so it's difficult to determine if a pig has it. Food manufacturers are required by law to heat pork products like ready-to-eat sausages to 137°F/58°C to kill the trichina.

While the practice of feeding garbage to pigs is declining in the US, it is still a low-cost option for pig farmers. Federal regulations require first heating the scraps to prevent the spread of diseases like trichinosis. Cases of trichinosis in humans are extremely rare (there are about 20 cases in the United States each year) and the risk of death is low, but it's still not something you want to take the chance of contracting.

KITCHEN TAKEAWAY Cooking pork to 140°F/60°C will kill any possible parasites, and the USDA recommends being even more diligent by cooking pork to an internal temperature of 145°F/63°C.

Is Grilled Meat Unhealthy to Eat?

THE ANSWER Maybe

THE SCIENCE We all love it—that tasty bit of charred crust that forms when meat is grilled over an open flame. But studies have found that there are two types of compounds found in charred meats that can increase the risk of cancer: heterocyclic amines (HCAs) and polycyclic aromatic hydrocarbons (PAHs). HCAs are formed when amino acids and sugars react at temperatures associated with charring, which are much higher than those that initiate the Maillard browning reaction. These temperatures can be reached when frying or grilling. PAHs are generated when fat and meat juices drip onto a very hot surface, catch fire, and create smoke. HCAs and PAHs are only really found in significant amounts in meat cooked by barbecuing, roasting, grilling, or other high temperature cooking techniques.

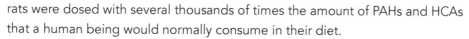

Cancer researchers have conducted studies to determine if PAHs and HCAs cause cancer by feeding these compounds to rodents that did, in fact, develop cancers and tumors in several different organs. However, the problem with these studies is that the rats were dosed with several thousands of times the amount of PAHs and HCAs that a human being would normally consume in their diet.

KITCHEN TAKEAWAY One good way to reduce exposure to PAHs and HCAs is to trim away any charred bits from grilled or barbecued food so you're not needlessly consuming these compounds. You can also trim away excess fat from steaks and chops before grilling them to minimize the chance of dripping fat hitting the heat source and generating PAHs.

DO I REALLY HAVE TO COOK MY STEAK TO WELL-DONE TO BE SAFE?

THE ANSWER No

THE SCIENCE A well-done steak has been heated to an internal temperature of 160°F/71°C or higher. That ensures that all bacteria have been killed off, inside and out. The trouble is that a well-done steak has been cooked to the point that it lacks juiciness, flavor, and tenderness. But is it really necessary to cook to those temperatures?

To get a better idea, let's first take a look at how beef is processed. The major bacteria concern when it comes to beef is *Escherichia coli (E. coli)*, which is found mostly in the cow's intestines. When cattle are slaughtered and processed, some of the bacteria from their intestines can get on the exterior surfaces of the cuts of meat, but none of the bacteria will actually seep into the steak. That's why a steak can be seared on the outside, remain at 130°F/54°C on the inside, and still be safe to eat—any bacteria that might be on the surface of the meat will be killed by the high searing temperature (*E. coli* are killed at 160°F/71°C). The trouble happens when steaks (like cube steak) are mechanically tenderized by puncturing them with sharp blades or needles, which contaminates the interior of the steak with bacteria-laden juices from the outside. Commercial ground beef is another potential hiding ground for *E. coli*, where contaminated surfaces are mixed together with the insides of the beef. Meat grinders are notoriously difficult to clean and can still harbor *E. coli* bacteria after deep manual scrubbing and sanitation. On the other hand, meat chunks that have had all their surfaces seared will not be contaminated with bacteria because none of the bacteria have made it into the interior of the meat itself.

KITCHEN TAKEAWAY The only time it's necessary to cook a steak to well-done is if the steak surface has been punctured by mechanical tenderizing. The interior of the meat is surprisingly sterile—so long as the surfaces have been seared above 160°F/71°C, there should no longer be any bacterial threat.

HOW DOES *E. COLI* END UP ON ROMAINE LETTUCE AND OTHER PRODUCE?

THE SCIENCE *Escherichia coli*, or *E. coli*, is a type of bacteria that is commonly found in the gastrointestinal tract of animals and humans. While most strains of *E. coli* are harmless, there are a handful that can cause infections in humans, the most common of which is *E. coli* O157:H7.

So how does produce end up getting contaminated with *E. coli*? Public health investigations have traced it to several different sources. One is when the feces of domestic livestock like cattle or pigs are washed into irrigation water and then find their way onto crops. Another is by exposure to improperly composted fertilizer containing animal feces.

Investigators have also linked contaminated produce with wild animals like feral pigs, deer, and birds that rummage through farm fields, graze on produce, and leave their feces on surrounding crops.

KITCHEN TAKEAWAY To avoid cross-contamination between food ingredients in your own kitchen, always wash your hands with soap (which kills bacteria) before handling different produce and meats. Cutting boards should also be washed with soap when switching between leafy greens and raw meats (or dedicate one board to meat and another to produce). *E. coli* and other bacteria like *Salmonella* can't be washed away with just water because they strongly adhere to food surfaces—cooking is the only way to get rid of them.

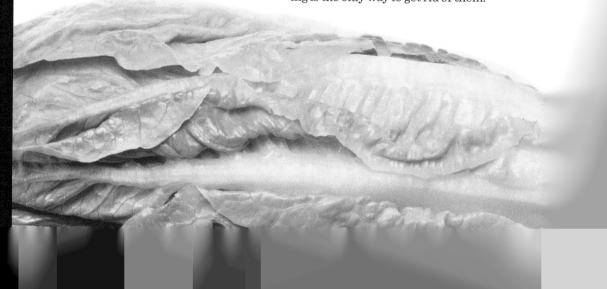

CAN THE DIRT ON MY MUSHROOMS MAKE ME SICK?

THE ANSWER No

THE SCIENCE Mushrooms are the fruiting bodies of a fungus, which is a large network of underground, thread-like fungal tissues known as mycelium. Mushrooms are grown commercially in two ways. Using the traditional, outdoor log-cultivated method, mushrooms are grown by inserting mycelium into holes that have been drilled into logs. The fungus is left to fruit mushrooms naturally or is forced to when the logs are soaked in cold water to mimic a spring or fall rainy season. This labor-intensive method produces high-quality mushrooms without dirt and is usually reserved for high-value medicinal or gourmet mushrooms, due to its irregular yield.

In indoor commercial mushroom farming, a special substrate made from a combination of sawdust, grain, straw, compost, and/or corn cobs is blended, placed in boxes or glass containers, and sterilized to remove contaminating microorganisms. The substrate is inoculated with fungal mycelium and allowed to grow in a carefully controlled environment. This is how most of the mushrooms you might buy in the supermarket are produced, and the dirt you may find on them is from this sterilized substrate.

KITCHEN TAKEAWAY Cleaning mushrooms individually with a dry paper towel or a special mushroom brush is time-consuming, and it's been shown in experiments that mushrooms absorb a minuscule amount of water when washed. It's fine to rinse them with cold water and dry them in a salad spinner.

Q WHAT KEEPS FERMENTED AND PICKLED VEGETABLES FROM GOING BAD?

THE SCIENCE The process of fermentation involves inoculating foods with special strains or cultures of bacteria, yeast, or molds to produce wonderful flavor while at the same time creating an environment that is unfriendly to organisms that are harmful to humans.

One of the most common bacteria strains used to ferment foods is lactic acid bacteria (*Lactobacillus*), which is used to ferment pickles, yogurt, kombucha, sour beer, and sauerkraut. Lactic acid bacteria metabolize the sugars in food and produce lactic acid. The presence of lactic acid reduces the pH of the food to about 3.5, a level of acidity that most bacteria cannot tolerate. Pickling works in the same way as fermentation does, except that, in this case, an external source of acid is added to the food, usually in the form of acetic acid (vinegar) or citric acid (lemon juice). Some strains of lactic acid bacteria will also release small concentrations of antimicrobial proteins that serve as natural preservatives and further inhibit the growth of competing bacteria.

Yeast, used to produce beer, wine, bread, and other fermented foods, creates an uninhabitable environment by converting sugars into ethanol, the alcohol found in beer and wine, which is quite toxic to many microorganisms. Some yeast strains can even produce and tolerate alcohol levels as high as 25 percent, in which very few microbes can survive.

Molds also produce natural compounds that are toxic to other microorganisms. Some of the molds used to ferment cheese belong to the genus *Penicillium*, also used to produce antibiotics like penicillin.

KITCHEN TAKEAWAY A safe salt brine for fermenting vegetables can be easily prepared by dissolving 2½ cups of kosher or pickling salt per gallon of water for a 10-percent starting brine. An additional ¾ cup of salt per gallon of water can be added for a final brine strength of 15 percent. The vegetables should then ferment over 4 to 8 weeks as sugars are converted to acid, keeping the fermented vegetables safe from other microorganisms.

Can I Still Eat a Potato If Its Skin is Green?

THE ANSWER Probably (without the skin)

THE SCIENCE If a potato is exposed to light, the skin will begin to turn green in preparation for photosynthesis as chlorophyll is produced and pumped into the surface of the potato. These processes also trigger the production of a toxic alkaloid called solanine, which can cause nausea, vomiting, diarrhea, stomach pain, and headaches if ingested in large quantities; 200 to 400 mg of solanine are required to intoxicate an adult, whereas only 20 to 40 mg of solanine will poison a child. A pound of green potatoes can contain between 10 and 65 mg of solanine; this toxin is concentrated in the shoots and skins of potatoes, so it's best not to eat potatoes that have sprouted or begun to turn green.

Potatoes are triggered into sprouting when they are stored in a warm environment, which simulates the coming of spring. Sprouting initiates the release of amylase enzymes, which convert the starches in the potato into sugars that feed the growing potato sprouts. The breakdown of the starches into sugar causes the potato to form wrinkles as water is osmotically drawn to the sugars.

KITCHEN TAKEAWAY Potatoes that have just begun to sprout are harmless and can be cooked after the sprouts are removed, so long as the potato is still firm. Potatoes that have become wrinkled or shriveled and have turned green should be thrown away. Potatoes that are still firm but whose skin is starting to look a bit green are also fine; just peel off and discard all of the green parts before using the potato.

ARE APPLE SEEDS AND PEACH PITS POISONOUS?

THE ANSWER Not really

THE SCIENCE Both apple seeds and peach pits contain a compound called amygdalin that breaks down into hydrogen cyanide during digestion. Cyanide is a known toxin with a fatal dose of 1.5 mg per kilogram body weight. However, the concentrations of amygdalin found in peach pits and apple seeds is such that a person weighing 137 pounds/62 kg would have to eat at least 235 peach pits or 875 apple seeds to receive a fatal dose. The human body can detoxify small quantities of hydrogen cyanide if just a few seeds are consumed. However, chronic exposure to moderate levels of cyanide over a long period can cause paralysis, headaches, nausea, and drowsiness.

KITCHEN TAKEAWAY Occasionally eating a few apple seeds (I'll assume you're not going to swallow a peach pit) isn't going to cause acute cyanide poisoning, but it's best not to eat them daily.

Do Castor Beans Contain Ricin?

THE ANSWER Yes

THE SCIENCE The castor bean is the seed of the castor bean plant and is used to produce castor oil, a common food additive and ingredient in food coatings. The interior of the castor bean also contains ricin, a protein that binds to carbohydrates. Ricin is highly toxic—the ingestion of 2 mg, little more than the equivalent of a few grains of salt, is enough to kill an average-size adult. Ricin works by inhibiting the synthesis of proteins inside cells, keeping them from carrying out their most basic metabolic functions.

But don't worry—because of the way it's produced, castor bean oil does not contain ricin. Lethal poisoning from ingesting castor beans is extremely rare because the seeds have a tough outer coating that makes them difficult to digest, and the acid in our stomachs deactivates most of the protein.

CAN PEANUT BUTTER GO BAD?

THE ANSWER Yes

THE SCIENCE The low moisture and high oil content of peanut butter prevents it from being spoiled by mold and bacteria. Few pathogenic bacteria can survive in peanut butter. However, peanut oil can become rancid. As soon as a jar of peanut butter is opened, fresh oxygen will come in and react with the oil, which is poly-unsaturated. Polyunsaturated oils are highly susceptible to oxidation, which forms peroxides, aldehydes, and other rancidity products over several months.

KITCHEN TAKEAWAY Expired peanut butter is not likely to make you sick, but the peanut butter oil can become rancid over time and make eating expired peanut butter an unpleasant experience. Rancid peanut butter tends to have a sharp, bitter, soapy, or metallic smell. When stored in the refrigerator, both hydrogenated and natural peanut butters can last twice as long as they will at room temperature, since the cooler temperature slows down the natural oxidation process. Whether it's stored in the refrigerator or at room temperature, natural peanut butter will become rancid faster than hydrogenated peanut butter because the hydrogenated version contains antioxidants that reduce oil oxidation.

DOES JAM EVER GO BAD?

THE ANSWER Yes

THE SCIENCE Commercial and homemade jams are kept preserved by their high acidity and sugar contents. As sweet as jam is, it has an acidity level (pH < 4.6) that is toxic to most spoilage microorganisms. Also, the high concentration of sugar in jam locks up moisture and prevents microorganisms from using it to metabolize nutrients. The shelf life of jam is further extended through pasteurization, which kills off any remaining microorganisms.

Once a jar of jam is opened, it is exposed to oxygen and a wide array of potential microbial contaminants. Again, most of these cannot make a home in jam, but there is a small group of yeasts and molds that have adapted to survive in high-sugar/high-acid conditions, so long as there is oxygen present. The most common of these is *Zygosaccharomyces rouxii*, a yeast that ferments to give soy sauce its distinctive flavor but that causes jam and other high-sugar foods to taste spoiled. Some species of *Penicillium* can attack jams even when they are refrigerated, producing mycotoxins that can affect the flavor.

KITCHEN TAKEAWAY Unopened commercial and homemade jams can last for up to 1 to 2 years when stored in a cool, dry place. Once a jam is opened, let your nose and eyes tell you whether it is spoiled. Telltale signs of spoilage are fermented, alcoholic, or yeasty odors. Light-colored jams will darken naturally over time, so darkening is not a sign of spoilage. Botulism in jam is rare because the acidity of commercial jams is enough to prevent the growth of bacteria that causes the botulinum toxin to form. However, homemade low-acid jams are at some risk of forming the botulinum toxin, so keep these jams refrigerated even when they are unopened and consume them as quickly as possible. (See page 156 for more on botulism.)

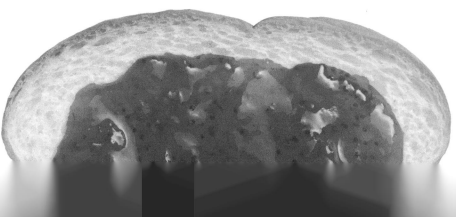

What Is Botulism?

THE SCIENCE Botulism occurs when food containing the potentially lethal botulinum toxin, produced by the bacteria *Clostridium botulinum,* is ingested. The botulinum toxin is one of the deadliest toxins produced by any bacteria and works by inhibiting nerve function, causing paralysis and respiratory failure. The good news is that *C. botulinum* will only grow and produce the toxin under very specific conditions, environments high in moisture and very low in or devoid of oxygen. Foods that have been properly dried or are exposed to air will not become contaminated with *C. botulinum;* it will also not grow at rates high enough to be toxic in foods with a pH below 4.6 and foods containing more than 5 percent salt, which is typical of most pickled foods. The types of foods that present a high risk for botulism are home-canned foods containing too little acid (a pH greater than 4.6) and jarred spices and foods in oil like roasted tomatoes, garlic, and rosemary.

KITCHEN TAKEAWAY Be careful if you choose to make your own flavored oils with fresh vegetables, herbs, or fruits, because that combination creates exactly the high-moisture, low-oxygen environment *C. botulinum* loves. The safest way to make flavored oils is to use dried herbs and spices, which won't introduce excess moisture to the oil. Flavored vinegars are relatively safe to make at home because vinegars have a pH below 4.6 and do not support the growth of *C. botulinum,* though they could potentially support the growth of acid-resistant *Escherichia coli.* To be safe, home-made vinegar infusions should be heated to boiling for 5 minutes to kill all bacteria.

 ## DOES OLIVE OIL EVER EXPIRE?

THE ANSWER Yes

THE SCIENCE There's an urban legend of sorts that high-quality olive oils age like wine. This is simply not true. Olive oil, no matter the type or price point, will break down over time as its acidity levels increase and its flavor weakens. But why does this happen?

Olive oil, like all oils, is made of molecules called triglycerides. These triglycerides have three long-chain tails composed of unsaturated fatty acids. The unsaturated designation refers to the way the carbons are bonded together in the chain. When olive oil is first cold-pressed, the triglycerides in the oil start off fresh and intact. But over time, olive oil begins to react with oxygen in the air to generate molecules called aldehydes and ketones. These molecules are responsible for the flavor and odor of rancid oil; if ingested in large quantities, they will make you sick. Water from the air will also interact with the triglycerides to create free fatty acids, which have an unpleasant taste.

KITCHEN TAKEAWAY You can use unopened olive oils within 18 to 24 months from the date of purchase; they will stay fresh for up to 2 months after opening. Store the olive oil away from heat and sunlight, which increase the rate at which oil breaks down. To test if olive oil has gone rancid, sniff it. Fresh olive oil should have a bright, grassy smell. If you don't use your olive oil on a regular basis, refrigeration will extend its shelf life.

Is It Safe to Reuse Oil After Frying and Cooking?

THE ANSWER Yes

THE SCIENCE If oil is properly strained and stored after cooking, it can be reused. However, reusing oil with water-containing food particles can create a breeding ground for *Clostridium botulinum* (see What Is Botulism?, page 156), due to oil's ability to trap water droplets and create an oxygen-free environment around them. The USDA recommends discarding unstrained used frying oil after one or two days; in that case, reheating the oil would kill the bacteria, but it may not necessarily destroy the botulin toxin the bacteria has produced. Another issue that can crop up when reusing oil is that the oil may be more prone to oxidation from exposure to air after multiple heating cycles. Oxidation can produce bitter flavors and slightly toxic compounds that can affect the human body. Oil that has acquired an off odor should never be used.

KITCHEN TAKEAWAY If you want to reuse frying oil, strain it through a fine-mesh strainer to remove food particles as soon as it has cooled down, and keep the oil refrigerated to slow down the growth of any potentially toxic bacteria. Refrigerated used oil is good for 1 month.

Q ARE HYDROGENATED OILS SAFE TO EAT?

THE ANSWER It depends

THE SCIENCE Hydrogenated oils are formed through a process called (not surprisingly) hydrogenation. When an oil is hydrogenated, its molecular structures become more rigid, allowing them to stack and lock into a solid at room temperature. This process allows manufacturers to convert liquid oils into hard fats, which add more creaminess and body to a food product. Hydrogenation also increases the shelf life and stability of these fats. The actual process involves combining the liquid oil with pressurized hydrogen gas at elevated temperatures in the presence of a finely dispersed metal powder (usually nickel, platinum, or palladium) as a catalyst, which speeds up the reaction. During the hydrogenation process, some of the oil molecules are only partially hydrogenated and re-form in a different way. These are known as trans fats. Trans fats also occur naturally but when they are consumed in high quantities, they can raise low-density lipoprotein cholesterol levels, which can lead to a buildup of cholesterol in arteries and raise the risk of coronary artery disease and stroke. The FDA ruled in 2015 that artificial trans fats were unsafe to eat and should be removed from the food supply by 2018. As a result, food manufacturers only use fully hydrogenated oils, which do not contain trans fats. These hydrogenated fats are fully saturated, do not have the same detrimental properties as trans fats, and have properties similar to natural saturated fats.

KITCHEN TAKEAWAY Fully hydrogenated oils function a lot like natural saturated fats and do not have the same health-deteriorating effects as partially hydrogenated oils. Currently, few food products contain trans fats, but they should still be avoided. Vegetable shortening, which is made from both partially and fully hydrogenated oils, used to be high in trans fats until Crisco introduced a low-trans-fat variety in 2004, and now all vegetable shortenings contain very little to no trans fats.

WHAT IS HIGH-FRUCTOSE CORN SYRUP, AND IS IT BAD FOR ME?

THE SCIENCE Corn syrup is a sugar syrup produced by breaking apart the starches found in corn. To make the syrup, corn is soaked in water, then milled into a powder. The powder is washed several times to dissolve the starches, which are separated out and then dried. Next, these starches are mixed with water and the enzyme amylase, which gets to work and breaks the starches into small carbohydrate fragments. Another enzyme, glucoamylase, is then added to the mash to break those fragments further down into glucose molecules. The crude glucose syrup is purified and exposed to the enzyme xylose isomerase, which converts most of the glucose sugars into fructose. The resulting product is high-fructose corn syrup, which is 42 percent fructose.

Fructose is almost entirely metabolized by the liver, where it's stored or used to produce fats, whereas glucose is metabolized by most organs and cells in the body, including the liver, red blood cells, brain, and muscles. Table sugar, which is sucrose, is metabolized by the body to form one molecule of glucose and one molecule of fructose. There's still a lot of scientific debate about how these differences in metabolism actually affect people in the long-term, but it's clear that many of the problems associated with high-fructose corn syrup have to do with the ingestion of excess calories and not necessarily the sugar itself.

KITCHEN TAKEAWAY Just like any other sugar, use and consume high-fructose corn syrup in moderation.

Is MSG Safe to Eat?

THE ANSWER Yes

THE SCIENCE MSG, or monosodium glutamate, is a naturally occurring amino acid and one of the compounds responsible for the taste we now call umami. Umami and its glutamate source were discovered by Japanese chemist Kikunae Ikeda, who also invented a process to produce glutamic acid, or glutamate. This process involved the acid hydrolysis of wheat gluten protein, which naturally contains very high levels of bound glutamates. The resulting glutamic acid was neutralized with sodium carbonate, yielding MSG, a more soluble form of glutamic acid. Today, that method has been superseded by a more economical approach that relies on a strain of bacteria, *Micrococcus glutamicus*, that is grown in fermentation vats filled with sugar and nutrients, much like brewing beer or fermenting wine. After several days of fermentation, the crude glutamate-rich broth is filtered, concentrated, crystallized, and converted to pure MSG.

In double-blind studies on the effects of MSG, researchers have found little difference in symptoms between individuals who consumed MSG-laden broths and placebo broths that do not contain MSG. Other studies found that when individuals who consumed the placebo broth were told it contained MSG or were asked if

THERE IS LITTLE SCIENTIFIC EVIDENCE
THAT ADDING MSG TO FOOD IS HARMFUL OR MAKES YOU SICK

they felt any side effects caused by MSG, they reported symptoms including headache, migraines, and chest pain, despite the fact that they did not actually consume any MSG. Additionally, people who consume glutamate-rich foods like cheese, tomato sauce, and mushrooms do not appear to report the same symptoms, despite the fact that these foods contain the same chemical component as MSG.

KITCHEN TAKEAWAY Glutamates are found naturally in high concentrations in a lot of different foods. There is little scientific evidence that adding MSG to food is harmful or makes you sick, so there's no reason to avoid using it. Dry soup mixes commonly include MSG, as do seasoning blends sold as flavor enhancers, like Ac'cent.

SHOULD I BE WORRIED ABOUT EATING FOODS CONTAINING ARTIFICIAL FLAVORS?

THE ANSWER   Possibly, but not in the way you think

THE SCIENCE   Natural flavors are extracted from plant or animal material and processed to improve their shelf life and performance in food products. Artificial flavors, on the other hand, are flavor molecules that have been chemically synthesized in the laboratory to have the same chemical structure as the main components of their natural counterparts. The reagents (the basic chemicals used for producing more complex chemicals) used in these processes are mostly sourced from petroleum and purified to meet rigorous regulatory standards. Indeed, artificial flavors undergo stricter safety evaluations than natural flavors do because of the perception that natural flavors are safer. However, many of the minor components of natural flavors have yet to be fully evaluated for their safety and could have unknown effects on human health. In either case, many of the same compounds are found in both natural and artificial flavors. So long as the chemical structures of those compounds are the same, whether from natural or artificial sources, they will elicit the same response from our brains.

An important difference between artificial and natural flavors is that natural flavors have more depth than their artificial counterparts. Take vanilla, for example. Vanilla extract from vanilla bean pods contains over 250 flavor and aroma compounds in addition to vanillin, its principal component. Imitation vanilla is mostly

VANILLA EXTRACT FROM VANILLA BEAN PODS CONTAINS OVER 250 FLAVOR AND AROMA COMPOUNDS

vanillin with a handful of other minor artificial flavors added to better mimic the natural version. The difference between the two when used in recipes and baked goods is subtle but noticeable.

A difference that argues in favor of artificial flavors is the ecological impact of using flavors derived from natural sources. Some, like the orchid that yields vanilla *(Vanilla planifolia)*, are grown in regions in the world that have fragile ecosystems. Using artificial flavorings helps to reduce the ecological pressure these regions experience.

KITCHEN TAKEAWAY There isn't much difference between natural and artificial flavors when it comes to your health. Flavorings should be chosen because of their flavor impact on food and not necessarily because of the perceived health benefits of their sources. Furthermore, the production of many artificial flavors has a lower environmental impact compared with the production of their natural counterparts.

Are Preservatives Safe to Eat?

THE ANSWER Yes

THE SCIENCE There are two types of common preservatives—antimicrobial preservatives and antioxidants—that serve different purposes in food.

Antimicrobial preservatives include sorbates, benzoates, nitrites and nitrates, and propionates; they are designed to prevent the proliferation of microorganisms like molds and bacteria that can produce toxins and cause illness. They function much like the way acetic acid in vinegar or lactic acid in yogurt keep microorganisms at bay by changing the pH of the food to an inhospitable level, except they're much more powerful at very low concentrations. For example, sorbates are used at a concentration between 0.025 percent to 0.1 percent, which translates to about 25 mg to 100 mg per 100 g of food. Benzoates are used in roughly the same range. Since these preservatives can be used at low concentrations, they do not affect the taste of the food. Studies conducted on the adverse effects of these preservatives found that they were minimal when consumed at these quantities. In contrast, natural toxins produced by molds and bacteria are known to cause severe allergies, cancer, organ failure, and even death if chronically consumed at low concentrations.

Antioxidants are used to keep foods from becoming rancid, producing off-flavors, and losing nutrients from exposure to oxygen. Some common antioxidants are ascorbic acid (also known as vitamin C), butylated hydroxytoluene, gallic acid, EDTA, sulfites, and tocopherols (compounds structurally related to vitamin E). Antioxidants work by either reacting directly with oxygen or scavenging metal ions that can catalyze the reaction between oxygen and chemical components of the food product. Again, many of these compounds have been thoroughly evaluated in animals and humans for their safety profiles. Some individuals may have allergies to sulfites.

KITCHEN TAKEAWAY Preservatives aren't toxins or poisons that should be feared. Their moderate consumption has been shown by the medical community to have few adverse effects. In fact, preservatives have helped reduce the incidence of many deadly food-borne diseases and improved the overall quality of many different types of food.

Q HOW SERIOUSLY SHOULD I TAKE EXPIRATION DATES?

THE SCIENCE A best-by date is supposed to represent the last date when the food will be at its highest quality and flavor; a sell-by date refers to the last date when a grocery store or retailer should sell the product; a freeze-by date is the last date before a food loses its peak quality unless it is frozen; and a use-by date is the last date before a food is expected to begin deteriorating or spoiling.

EXPIRATION LABELING IS NOT REQUIRED
BY FEDERAL REGULATION

There are also sometimes codes associated with these dates, which are used by manufacturers to identify the date and time of a food's production. There's no uniform structure for these codes, but months are usually designated by a letter (A through L represent January through December) or a number and/or letter (1 through 9 represent January through September, and the letters O, N, and D represent October, November, and December). The year and day are often posted as a string of numbers and/or letters.

Expiration labeling is not required by federal regulation (except for on infant formula); it is up to the food producer's discretion. These dates are not indicators of product safety. Food should still be safe if properly handled and stored past these dates. You, the consumer, are the best judge of whether a food has gone bad: Signs of spoilage include off odors, unpleasant flavors, changes in texture, and the presence of mold. Cans that are bulging, swelling, or have dents can harbor the *Clostridium botulinum* bacteria that forms the toxin responsible for botulism and should be avoided. Damage can create microscopic holes in cans that allow small amounts of bacteria into a very low-oxygen environment, which allows *C. botulinum* to grow and produce its toxins.

KITCHEN TAKEAWAY You can still eat foods beyond their expiration dates, as these dates are manufacturer estimates associated with product quality and are not associated with food product safety. Just be sure to recognize the indicators of food spoilage, like changes in color or texture, unpleasant odors, or undesirable tastes, and handle and store foods properly.

Do I Need to Store Soy Sauce or Fish Sauce in the Fridge Once I Open It?

THE ANSWER No

THE SCIENCE Soy sauce and fish sauce are made by fermenting soybeans or fish with enzymes to break down the proteins into amino acids. In the case of soy sauce, this is done with enzymes produced by the mold *Aspergillus oryzae*. For fish sauce, the fish tissue itself contains enzymes that are released as the fish decomposes. In both cases, they are mixed with copious amounts of salt or salt brine to protect the fermenting sauce from infection by unwanted microorganisms. Many pathogenic microbes are sensitive to salt and will dehydrate in high salt environments. As these sauces age, they generate more acid, a by-product of the fermentation process. After several months to several years of fermenting, the sauce is strained and pasteurized to stop the fermentation process. Because of the high salt content and high acidity of both these products, there is no danger of microorganisms

making their home there, whether stored at room temperature or in the fridge. These are two ingredients that will keep pretty much forever. Other ingredients that don't need to be refrigerated are vinegar, hot sauce, oyster sauce, and honey.

KITCHEN TAKEAWAY Feel free to leave your soy sauce or fish sauce in the cupboard. There's no danger of any nasty microbes making a home in these sauces, but you can place them in the fridge to help preserve their flavor.

SHOULD I STORE MY BREAD IN THE FRIDGE?

THE ANSWER No

THE SCIENCE The starch granules in wheat grain are largely in a crystallized form. When wheat grain is pulverized into flour, mixed with water to form dough, and baked in the oven, the crystallized starches become hydrated and lose their orderly shape, becoming amorphous and gelatinized; this is what gives bread its softness. However, this amorphous state is only semi-stable, and the starches will slowly return to crystallized form over time through retrogradation and recrystallization. Because crystalline starches are packed more tightly than amorphous starches, water is forced out. The result is the loss of moisture and eventual staling that gives stale bread its hardness.

At room temperature, the rates at which these processes occur is slow, but when bread is placed in the refrigerator and cooled down, the lower temperature causes the bread starches to revert to their original crystalline form at a much faster rate.

KITCHEN TAKEAWAY For long-term storage to prevent mold and ensure the best texture, it's a better idea to keep your bread in the freezer and defrost slices as you need them. At freezing temperatures, retrogradation is slowed down because the water molecules are locked in as ice. Otherwise, leave your bread out at room temperature if you plan to eat it quickly.

KEEP YOUR BREAD IN THE FREEZER
AND TAKE OUT SLICES TO WARM TO ROOM TEMPERATURE

CAN I JUST CUT THE MOLDY BITS
OFF BREAD AND STILL EAT IT?

THE ANSWER Maybe, but honestly, not such a great idea

THE SCIENCE Mold is a type of fungus that spreads when its spores, which are naturally present in the environment, take root on food and begin germinating. Mold spores are highly durable and resistant to heat and dry conditions, so they can be found nearly everywhere. Like most organisms, mold grows through cell division. What's unique to molds is that their cells can merge to form large networks of cellular filaments with multiple nuclei, called mycelium, which allow molds to deeply penetrate a food. This means that a patch of mold on a piece of bread may have penetrated other parts of the bread without being visible.

All molds secrete enzymes into their surrounding environment to digest the proteins and carbohydrates available to them, producing simpler sugars and amino acids, which they absorb (and which is why moldy foods tend to soften over time). To help maintain their microbial dominance, molds also produce compounds called mycotoxins, which are poisonous to other microorganisms, as well as to humans. Several species of molds grow on bread, the three most common being *Rhizopus stolonifer*, or common black bread mold; mold species from the genus *Penicillium*; and mold species from the genus *Cladosporium*. Black bread mold does not usually produce mycotoxins, but under certain conditions, some species of *Penicillium* and *Cladosporium* can produce mycotoxins that can trigger an allergic reaction and respiratory problems in some people.

KITCHEN TAKEAWAY There's little reason to eat moldy bread. It's best to throw it away to be on the safe side, since that spot of mold could be the tip of an invisible iceberg that could cause you a world of hurt for a few hours.

SHOULD I KEEP UNRIPE TOMATOES IN THE FRIDGE?

THE ANSWER No

THE SCIENCE Tomatoes contain a set of enzymes that generate the compounds responsible for the ripened fruits' unmistakable aromas and flavors. When unripened tomatoes are placed in the fridge, the low temperatures cause these aroma- and flavor-forming enzymes to shut down. These enzymes can be reactivated in tomatoes that have been stored in the fridge for one to three days; however, after three days in the fridge, the tomatoes lose their ability to produce flavor compounds, which is why most supermarket tomatoes are pale in color and flavor.

Picking unripe tomatoes cuts them off from the tomato vine's supply of sugars and other nutrients. Only green tomatoes that have a faint red or yellow blush at the bottom will continue to ripen off the vine, because they already contain the nutrients necessary to continue developing. Unripened tomatoes also produce ethylene gas, which aids in ripening, but the ethylene does not cause them to ripen as well as they would on the plant. Although putting a banana or other ethylene-producing fruit (see page 96) together with a completely green tomato may make it turn red, that method will not necessarily yield the ripened flavor we crave; flavor development in tomatoes requires more time than ethylene-accelerated ripening can offer.

KITCHEN TAKEAWAY Unripened tomatoes should be kept at room temperature away from sunlight and allowed to ripen before using. Fully ripened tomatoes at the peak of their flavor should be eaten right away, or they can be refrigerated just a day or two; they will lose their flavor if refrigerated for more than three days. If you have refrigerated a tomato, let it come back to room temperature before using it for the best flavor.

WHY DO VEGETABLES
WILT IN THE FRIDGE?

THE SCIENCE After just a few days in the fridge, you might discover that your carrots or celery stalks have begun to soften or wilt. Certainly, all fruits and vegetables have a life span, at which point they wilt and then rot. But more often than not, the reason why your produce has wilted is because it has lost water. Produce have tiny pores in their skin called stomata that allow them to breathe. They will continue to breathe oxygen long after harvesting. Water can evaporate through these pores, which is normally replenished by the roots of the original plant from which they're grown. When fruits and vegetables are placed in the fridge, they start to lose moisture quickly because modern refrigerators are designed to maintain low humidity to keep food dry. As a result of this loss of moisture, the plant cells collapse and the produce wilts.

KITCHEN TAKEAWAY The easiest way to restore produce back to its natural crispness is to soak it in water in the refrigerator. Just like wilted flowers, wilted fruits and vegetables will firm up after a nice drink of water.

Q HOW DOES A FREEZER WORK?

THE SCIENCE Freezers work through an electric pump that compresses a gaseous refrigeration coolant into a liquid. Driven by the electric pump, fast-moving gas molecules become concentrated into a smaller volume of hot liquid. This newly formed hot liquid is then pumped through a series of condenser coils and the heat is radiated out to the surrounding environment, which is why the back of a freezer is so hot. As the liquid cools down to room temperature, it gets pumped through an expansion valve and back into a series of evaporator coils that flow into the interior of the refrigerator or freezer. The evaporator coils have a lower pressure than the condenser coils, which allows the coolant liquid to evaporate and expand back into a gas. The net result is that the bulk of the liquid cools down, and heat from the freezer compartment is carried away by the rapidly expanding gas. The warm gas is than re-pumped back into the condenser coils and the process is repeated in an (almost) endless loop. To save on energy, freezers are operated in cycles. When the temperature of the freezer rises to a certain point, the pump kicks in and restarts the process to cool down the freezer. Opening the freezer door can also trigger the process into action.

THE FREEZER TEMPERATURE SHOULD BE SET TO
0°F/−18°C

KITCHEN TAKEAWAY Freezers and refrigerators should be kept at their proper temperatures. Set the refrigerator temperature at 35°F to 40°F/~2° to 4°C, which is the range that slows down the growth of bacteria without freezing the food. The freezer temperature should be set to 0°F/–18°C. While lower freezer temperatures than 0°F/–18°C can freeze food faster, setting your freezer to that temperature will lead to a higher than necessary electricity bill. Temperatures should be checked periodically to make sure your appliance is working properly.

How Your Freezer Works

The cooling power of the freezer (and refrigerator) relies on a cycle of compressing and expanding coolant through a series of pumps, valves, and coil. Ultimately, cold air is generated, as well as hot air.

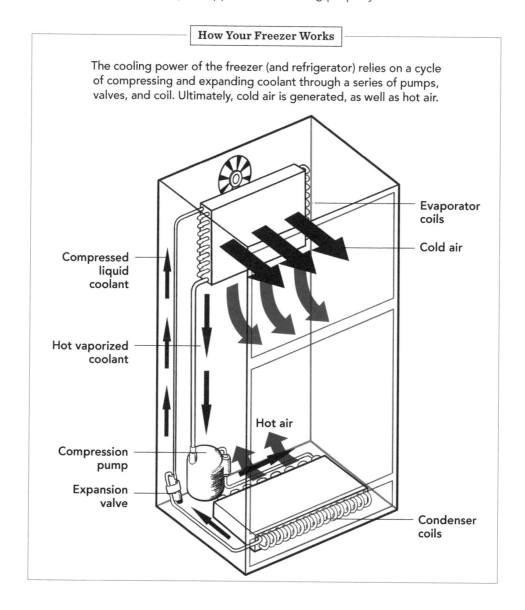

Compressed liquid coolant

Hot vaporized coolant

Compression pump

Expansion valve

Evaporator coils

Cold air

Hot air

Condenser coils

Why Do You Have to Blanch Green Beans Before Freezing Them?

THE SCIENCE Normally, the nutrients and enzymes in vegetables are compartmentalized inside the plant's cells and kept separate from one another. However, as vegetables freeze, the water in the cells expands and breaks open the cell walls, exposing the cell contents. The enzymes that are contained within the cells can affect the flavor, texture, and color of certain vegetables, degrading their quality and shortening their shelf life. While enzyme reactions are very slow at freezer temperatures, they can still occur and affect the quality of the vegetables over time.

Blanching vegetables briefly in boiling water deactivates these enzymes and can extend the freezer life of produce from 2 months (for unblanched vegetables) to 12 months. In addition to green beans, asparagus, broccoli, and okra also benefit from blanching before freezing. Some vegetables, such as bell peppers, onions, corn, and tomatoes, do not need to be blanched before freezing because they contain significantly fewer of the enzymes that cause the cell walls to fall apart and soften the vegetable during storage.

KITCHEN TAKEAWAY To blanch vegetables, place them in a large pot of boiling water (salted, if desired) and cook until the vegetables are crisp-tender, about 2 to 5 minutes. Then, quickly drain the vegetables and drop them into a large bowl of ice water to stop the cooking. Drain the vegetables one more time, patting them dry with clean kitchen towels before freezing to get rid of any water clinging to their surfaces.

Q WHAT IS FREEZER BURN?

THE SCIENCE Though you know that a cup of water will eventually evaporate on a warm day, it may come as a surprise to learn that ice can also evaporate, even at freezing temperatures. Known as sublimation, the process is quite slow, but if the air is sufficiently dry, ice can vaporize and migrate from one material to another. This is what happens when foods experience freezer burn, which is basically the process of frozen foods drying out in the freezer.

When food loses moisture, it loses frozen water molecules that would normally form a dense, crystalline barrier around the proteins, fats, and carbohydrates that make up the food and serve as a protective layer between them and oxygen. As layers of water are stripped away, oxygen in the freezer can start to react with the fats, pigments, and flavor molecules in food, transforming them into intensely disagreeable molecules. The result is discolored, funky-tasting food made that way not by microbial spoilage but by dryness.

If a food is completely dry (say, toasted bread crumbs), freezer burn won't occur because there isn't enough water to act as a catalyst. The moisture level of freezer-burnt food lies in that Goldilocks zone where there's just enough water that oxygen can do maximum damage.

KITCHEN TAKEAWAY Throw away any food that exhibits freezer burn—it's beyond salvation. To help prevent freezer burn, wrap your food tightly in waterproof packaging so moisture can't escape and air can't easily come in contact with the food.

OXYGEN IN THE FREEZER
CAN START TO REACT WITH
THE FATS, PIGMENTS, AND FLAVOR
MOLECULES IN FOOD

WHY DOES ICE CREAM SOMETIMES DEVELOP ICE CRYSTALS IN THE FREEZER?

THE SCIENCE Ice cream is a complex emulsion of air, water, fat globules, and sugar, typically made by simmering sugar, cream, and milk together and churning the mixture in an ice cream maker. Ice cream makers are just some kind of freezing bowl or freezing machine that churns and cools the mixture down below freezing. During the churning process, air bubbles are whipped into the slowly freezing mixture while ice crystals begin to grow, creating that delectable texture that we love. The key to great ice cream texture is for very tiny ice crystals to form quickly and remain uniformly dispersed throughout the mixture; commercial manufacturers add emulsifiers and stabilizers to aid in this.

However, those tiny ice crystals can go through a process called Ostwald ripening if exposed to temperature fluctuations, where they slowly dissolve and then re-form into larger crystals. It's the same idea as a vinaigrette separating into its oil and vinegar constituents (the individual oil droplets break out of the emulsion and join with one another until they've fully separated from the vinegar). These larger ice crystals give ice cream that coarse, gritty texture that signifies the treat has been left in the freezer for too long.

KITCHEN TAKEAWAY Ostwald ripening is thermodynamically driven; to keep it from happening (or at least minimize the possibility), keep your ice cream in the part of the freezer with the most stable temperature, like the back. Ice cream is more likely to develop freezer burn if it's kept in the door or near the front of the freezer.

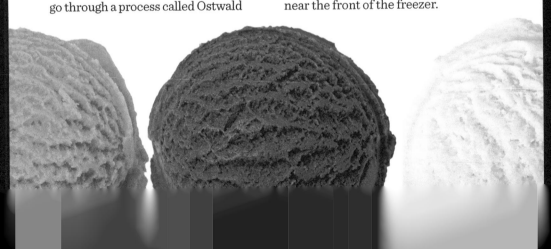

What's the Best Way to Defrost Meat—on the Counter, in the Fridge, or Under Running Water?

THE SCIENCE At temperatures between 40°F and 140°F/4°C and 60°C, also known as the danger zone, microbes can proliferate rapidly. The USDA recommends that meat be kept at or under 40°F/4°C to ensure that no bacteria populations take hold. The best way to maintain those temperatures is to place frozen meat in the refrigerator to defrost completely. That way, the meat never climbs into the danger zone's temperature range.

When meat is left on the counter or placed under running water to defrost, it does so unevenly. The meat may still be icy in the middle while the exterior becomes warmer from its surroundings. That's where bacterial growth can occur. For the most part, spoilage or pathogenic bacteria populations will still be low if the meat is kept on the cooler side of the danger zone. However, if the meat was improperly handled, stored, or transported before it reached the grocery store display, the risk of these bacteria proliferating when the meat is defrosted outside the refrigerator will increase.

KITCHEN TAKEAWAY Plan ahead as best as you can to safely defrost meat by moving frozen meat into the fridge a day or more ahead of when you need it. For every 5 pounds of meat, factor in 36 hours of thawing time in the refrigerator. Even small amounts of meat can take a whole day to defrost, and steaks, ground beef, chicken breasts, and stew meat can take between 24 and 48 hours to thaw completely.

PLAN AHEAD
AS BEST YOU CAN
TO SAFELY DEFROST MEAT

Index

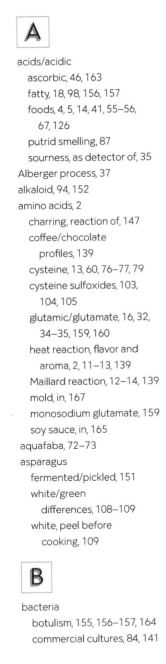
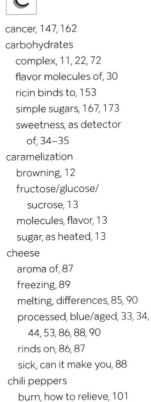

umami

Acknowledgments

THIS BOOK PARTIALLY CAME OUT OF MY TIME spent writing about food science and technology for the Institute of Food Technologists Student Association's *Science Meets Food* blog. I am grateful to my team at *Science Meets Food*, whose passion for writing, food, and science inspires me every day, and I'm thankful for the IFTSA Board of Directors for giving me the chance to help run the organization's blog and social media as the association's VP of digital and social media.

I want to include a special note of thanks to my wife, Yvonne Le, for her willingness to read through and critique my first drafts, listen to me drone on about the differences between onions and garlic, and endure years of endless experiments in both the kitchen and the lab.

Of course, I want to thank my mom for letting me build a makeshift laboratory in the garage and spend hours purifying chemicals from household cleaners as a kid. And I'd like to thank my family for watching me perform chemistry magic shows during Christmas holidays.

I want to also thank my editor, Pam Kingsley, for her sharp eyes, professional acumen, insightful questions, and for the time she spent reading through my drafts. Finally, I want to thank the team at Callisto Media for giving me this incredible opportunity to write my first book on a topic I am so passionate about!

About the Author

BRYAN LE is a PhD candidate in Food Science at the University of Wisconsin, Madison. He serves as the VP of digital and social media for the Institute of Food Technologists Student Association and manages, edits, and writes articles for their official blog, *Science Meets Food*. His publications have been featured on Medium, *Heated*, *Technology Networks*, and the Kerry Health & Nutrition Institute. Le is also a recipient of the James Beard Legacy Scholarship from the James Beard Foundation. In another life, he walked 2,000 miles from Los Angeles to New Orleans. He is an avid runner and currently lives in Madison, Wisconsin, with his wife. For more information, visit him at BryanQuocLe.com.